LOCAL SUSTAINABILITY

Paul Selman has worked in planning, and has lectured on environmental planning and management in various universities and colleges. He is Professor of Environmental Planning, Cheltenham and Gloucester College. He has published extensively and is editor of *Landscape Research*, and his books include *Environmental Planning* (1992), *Countryside Planning in Practice* (ed) (1988) and *Ecology and Planning* (1981).

LOCAL SUSTAINABILITY

MANAGING AND PLANNING
ECOLOGICALLY SOUND PLACES

Paul Selman

UNIVERSITY OF WOLVERHAMPTON
LIBRARY

Acc No.	CLASS 333
2095912	333.7
CONTROL	
1853963003	Se
DATE 27 JUN 1997	SITE WV

P·C·P
Paul Chapman
Publishing Ltd

Copyright © 1996, Paul Selman

All rights reserved

Paul Chapman Publishing Ltd
144 Liverpool Road
London
N1 1LA

Apart from any fair dealing for the purposes of research or
private study, or criticism or review, as permitted under the
Copyright, Designs and Patents Act, 1988, this publication may be
reproduced, stored or transmitted, in any form or by any means,
only with the prior permission in writing of the publishers, or
in the case of reprographic reproduction, in accordance with the
terms of licences issued by the Copyright Licensing Agency.
Inquiries concerning reproduction outside those terms should be
sent to the publishers at the abovementioned address.

British Library Cataloguing in Publication Data

Selman, Paul H. (Paul Harry)
   Local sustainability
   1. Sustainable development
   I. Title
   333.7

   ISBN 1 85396 300 3

Typeset by Dorwyn Ltd, Rowlands Castle, Hants.
Printed and bound in Great Britain

A B C D 9 8 7 6

Contents

Preface

Sustainable development has become one of the catch phrases of the 1990s. As a term, it merits close study if only because of the huge number of official reports and academic papers it has inspired. Most of the texts on the subject have tended to emphasise theoretical or global aspects. My original intention was, therefore, to pull together some of the literature relating to the sub-national levels of policy and practice, principally from a western perspective. This was particularly influenced by my training as a town planner and, essentially, my initial project was to write a text on the way in which the planning profession could contribute to sustainable development. Once started, it became clear that the role of planning departments could not easily be separated from the wider work of local authorities, and so the scope rapidly broadened. This was further compounded when, after a while, I became increasingly unhappy about equating environmental practices in local government with the governance of local sustainability. The two were interconnected, but not synonymous. This led to the eventual focus of the book, namely the convergence of concerns and interest groups at the local level which might be conducive to the pursuit of sustainability.

Given that this book was produced within the interstices of one very busy year, I frequently found myself going up an almost vertical learning curve. I hope, therefore, that what my critique lacks in finesse, it makes up for in freshness. If nothing else, it confirms that sustainable development cannot be achieved by one group of professionals acting in an official capacity and implementing a standard solution. It is a transdisciplinary and multi-professional concept, which must involve the whole community in a risky discourse. Above all, on finishing the manuscript I found that the importance of sustainability had increased in my estimation rather than, as I had initially anticipated, dwindled. I hope that I have managed to convey some of my own fascination with the complexities and potentials of the subject, and my firm conviction that its urgency and relevance remain undiminished.

Paul Selman

Acknowledgements

Some of the material in this book draws on the research conducted under a grant from the Economic and Social Research Council, 'Policy, Process and Product in Local Agenda 21', grant number L320253221.

The author and publishers acknowledge the following permissions to reproduce copyright material :
Figure 1.1 and Table 2.1 from J Carew-Reid et al (1994) *Strategies for National Sustainable Development: a handbook for their planning and implementation*, Earthscan, London, in association with the IIED and IUCN.
Figure 3.1 from A-M Jansson, H Hammer, C Folke, R Constanza (eds) (1994) *Investing in Natural Capital; the ecological economics approach to sustainability*, Island Press, Washington D.C.
Figure 3.2 and Table 3.3 from H van der Vegt, H ter Heide, S Tjallingii and D van Alphen (eds) (1994) *Sustainable urban development: research and experiments*, Delft University Press.
Figures 3.3 and 7.3 from A Blowers (ed) *Planning for a Sustainable Environment* (1993), Earthscan, London.
Table 4.7 reprinted with permission of the International Association for Landscape Ecology.
Figure 4.2 from S Roebuck and A Gurney, *Planning Week*, Vol 3 no 22 p 17, reprinted with permission of *Planning Week*.
Figure 4.5 from G. O'Brien and C N Gibbins (1994) Conducting a Preliminary Environmental Review : a case study of Newcastle International Airport, *Environmental Policy and Practice*, Vol 3 no 4 p 267, reprinted with permission fiom EP Publications.
Figures 5.1 and 5.2 from *A Guide to the Eco-Management and Audit Scheme for UK Local Government: a manual for environmental management in local Government* (1993) HMSO. Crown Copyright reproduced with the permission of the Controller of HMSO.
Table 6.4 from P Selman, (1995) Local Sustainability : can the planning system help us get from here to there? *Town Planning Review*, 66 (3) p 299. Reprinted with permission from *Town Planning Review*.
Table 7.1 reprinted with permission from the ICC United Kingdom.

1

The nature of local sustainability

Introduction

The idea that the quantity and quality of the Earth's resources are scarce relative to demand is not new. Nor is the notion that people ought to be involved in solutions for the planning and management of their own area. Yet terms such as 'sustainability' and 'sustainable development' are relatively novel and have been associated with radical new agendas. Along with their rapid rise to prominence has been a recognition of the scope for local, often lay, involvement in devising and implementing sustainable development solutions. The importance which has been attached to local people and their representative councils is in sharp contrast to the centralised, expert-led approaches which have dominated twentieth-century practice. Equally striking has been the way in which the sustainability agenda has has led to the integration of environmental, social and economic issues, as opposed to their traditional separation into distinct policy areas.

Many societies have long practised principles of stewardship in their management of farmland, forests and fisheries. They have been well aware that sustainable yields are only achievable in the absence of unchecked greed. Even with regard to non-productive, amenity resources, such as scenic or touristic assets, there has been a long-standing recognition of their vulnerability and importance. The desire to protect and sustain 'cultural' landscapes, including their aesthetic, historical and recreational qualities, dates back at least to the eighteenth century. These varied concerns reflect the need to protect local environmental resources, and to ensure that they are used 'sustainably'.

Yet the explicit adoption of sustainability as a touchstone of urban and rural development has been a very recent phenomenon. This is a consequence of the gradual accretion of scientific evidence about the excessive exploitation of the earth's biological and physical resources, and a growing acceptance that irreversible environmental disasters pose a palpable threat. Also, we have begun to recognise that natural resources possess more than those merely utilitarian values which satisfy the need for human survival. Particular resources may be worth sustaining for a variety of economic, scientific, aesthetic, symbolic and even spiritual reasons. This greatly magnifies the need for sensitive and carefully negotiated use of the environment.

Whilst the early stages of the environmental revolution (in the 1960s and 1970s) focused on the damage inflicted on natural systems, the more mature phase (in the 1980s and 1990s) has given greater prominence to the role of people. Thus, initial explanations of environmental damage pointed mainly to the physical demands placed on land and water resources by crowding and consumption. Latterly, these accounts have been partially supplanted by analyses which recognise the damage caused by ill-conceived, top-down solutions to economic development, and by the inequalities associated with capital restructuring. In both developed and developing countries, the production of plans, programmes and projects purely by 'experts' has been associated with unsustainable use of natural resources and 'boom and bust' investment. Indigenous solutions, capturing locally based enthusiasm and knowledge, are now often deemed more appropriate.

The celebration of the local and indigenous is, however, an incomplete analysis. It under-estimates the difficulties of defining and implementing local solutions, and tends to ignore the distortions of local economies by powerful external pressures. The ways in which urban and rural societies are being restructured by corporate interests has been widely discussed elsewhere. The intention here is to examine the ways in which actions for sustainability may be harnessed within the local arena, in contrast to the national and global orientation of the traditional environmental agenda.

Perhaps most significant in promoting the notion of sustainability has been the recognition of the human dimension. For example, the importance of (local) people is exemplified in the many serious and irreversible instances of bad environmental management which have arisen through poverty or unjust trading relations. More generally, there is an acknowledgement that we cannot simply depend on governments to take action, nor can state organisations merely instruct individuals to be more environmentally responsible. Citizens and local communities have to believe in the need to adopt more sustainable lifestyles, and act accordingly. Whilst the three decades preceding the 1990s saw a progressive rise in collective national and international responses to environmental pressures, the closing years of this century are witnessing a shift towards local responsibility in attaining sustainability. The need is thus becoming one of reactivating citizenship and community pride, and encouraging them to work in tandem with wider environmental initiatives.

Thus, whilst most of the environmental literature has dwelt on the broad framework of resource depletion and institutional responses, the environmentalist's gaze is now shifting to the dilemmas posed, and the human energies available, at the local level. It is here that the tough choices associated with a transition to sustainable development are frequently cast into sharpest relief.

People's principles are put directly to the test when they are confronted by actual proposals as opposed to hypothetical and abstract planning options. This means that at the local level, the intrinsic differences between apparently concordant environmental groups can manifest themselves as conflict.

In developed countries, where amenity is a prominent issue, local choices might include balancing the landownership rights of producer groups (farmers, foresters, etc.) with the less clearly defined user rights of people who wish to access the same land for recreation or conservation purposes. Or local residents who welcome a new road proposal because it will relieve them of traffic and noise may be opposed by local (or outside) preservation groups who do not wish to see land sacrificed to a highway. In both developed and developing countries, choices may relate to a deeper conflict between economic survival and environmental ethics. For example, conservationists (both locally and distantly based) may be in conflict with local land managers or fishers by trying to prevent them improving their land or culling wild species. Some dubious actions may have been undertaken by a branch factory or farm which is part of a wider international commercial operation, so that the local area can become the arena in which a wider conflict is played out.

These local conflicts are often complicated by imbalanced power relations between opposing camps, with some parties (such as developers, state departments and transnational corporations) wielding greater influence than citizens or voluntary bodies. Conflicts which are apparently local and associated with specific land use proposals may therefore be highly complex and relate to matters of national or international significance. The acronym NIMBY ('not in my back yard') has often been used pejoratively to dismiss local opponents of apparently desirable schemes by making them appear ignorant and selfish. Yet their opposition, though perhaps poorly articulated, may be a legitimate response to externally generated demands on local assets. Some political analysts have, indeed, interpreted it as dissent against the colonisation of the local arena by the central state.

All this suggests that, far from being the unglamorous relation of its global counterpart, the local 'patch' is often the crucial arena in which sustainability must be pursued. It is where conflicts arise, attitudes are changed and actions are instigated and hence, where the need for locally-generated solutions to the problem of achieving a sustainable use of the environment is increasingly recognised. Sustainability will not be achieved by edict, but rather by the conscious reconstruction of our policies, institutions, attitudes and actions at all scales from the global to the very local.

Local focus

Clearly, the role of the locality is pivotal in moving towards an environmentally sustainable future. The notion of 'locality' is twofold. In physical terms, it refers to a place, to the people who inhabit it and to the spatial and administrative systems around which their activities revolve. In social terms, it represents groups of individuals who are associated through common responsibilities, occupations, cultures or interests (c.f. Lowe and Murdoch, 1993). Frequently, there is a considerable degree of geographical correspondence between these two interpretations of locality. The physical concept is important in setting a framework for measuring, monitoring and managing environmental resources within the context of local administrative units. The social definition underpins our attempts to mobilise citizens and their constituencies of interest, together with the networks and partnerships within which they operate at the local level.

Whilst international bodies and national governments must continue to be prime movers in reconciling economic and ecological objectives, their statements will be pious and empty if they fail to convince individual citizens, households, businesses and organisations. The former 'macro' dimension is important for securing international co-operation, for law- and policy-making, and for instigating debate and research on new areas of environmental science, economics and ethics while the latter 'micro' dimension is critical to the generation of an enduring commitment to environmental responsibility.

The other side to this coin is that, if the local voice is to be influential, it should seek to achieve more than a veneer of sustainability. It is easy for a community to believe that it has 'greened' itself by making more use of recycling facilities and planting more trees, at the same time as it continues to be grossly profligate in resource use generally. Even if a locality is unable to take immediate direct action to reduce this impact, residents should at least raise their level of awareness of the magnitude and nature of their demands for materials and energy, and their export of pollution and wastes, which affect other locations. This does not imply that localities should generally pursue a path of self-sufficiency. Rather, they should become progressively as frugal as is consistent with a reasonable lifestyle, and should not propel other areas onto an unsustainable trajectory. In order to make this transition they will require reliable information and feedback, practical support and encouragement, and a strong sense of citizenship.

A key player in the transition to sustainability will be local government. Despite experiencing a significant erosion of its powers in the UK, local government is nonetheless pivotal to the representation of people's interests in their locality, and to the provision of local leadership. Local authorities, it has been said, are important precisely because they are local and have authority. At a formal level, local government affects the environment directly, through its use of resources such as recycled paper and energy; it also causes impacts on the environment indirectly through its policies which relate to its administrative area. Some of its most important policies are those derived from the town and country planning system which provides a negotiated basis for allocating environmental resources. Parallel to the statutory planning system has been the recent emergence of a raft of non-statutory 'green planning' mechanisms which report environmental conditions, support policy implementation, and audit performance. Frequently, local government will be the instigator of sustainability measures within its area, and its role features strongly in the subsequent text. However, increasingly its test of character is seen to be the degree to which it is willing to concede environmental powers and responsibilities to other stakeholders.

The nature of sustainability

The concept of sustainability is not, in essence, particularly novel. The principle of 'sustained yield' has been understood for many decades, especially in relation to fisheries and forests. More general ideas of non-destructive resource management have permeated indigenous cultures for very much longer. However, its popularisation as a widespread term has been much more recent. Perhaps the most significant origin of its usage was the World Conservation Strategy (International Union for the Conservation of Nature and Natural Resources, 1980), whose central tenet was *resource conservation for sustainable development.*

Up to that point there had been a strong tendency to portray conservation as a luxury which only the wealthier nations could afford and it was closely associated with touristic and residential amenities. Even so, it was only tolerated as long as it did not interfere with development opportunities. It was widely assumed that conservation priorities could reasonably be overridden where they were in apparent conflict with economic growth and human welfare. The WCS was a watershed in challenging the idea that conservation and development were intrinsically opposed, or even that they could be separated.

In practice, the prevailing view that development and conservation were polarised and opposing forces was disastrous for effective environmental management. The continued attrition of soils, seas and genetic resources across most parts of the earth, all of which are fundamental to sustained human wellbeing, were clearly dependent on effective conservation. At the same time, it was gradually becoming apparent that many capital-intensive projects, especially though not exclusively in developing countries, had been counterproductive in their effects. Hailed as growth points for regional economies, they had often left unemployment, social dislocation and environmental devastation in their wake. The WCS succeeded in shifting conservation to centre stage as a fundamental prerequisite to economic development: continued human welfare was portrayed as being firmly founded on the inherent capacities and indigenous ecological and social capabilities of the host locality.

Naturally enough, whilst the principles of sustainable development appeared compelling, they did not necessarily lead to harmony and consensus on the resolution of individual cases. Indeed, the notion of sustainable development opened a veritable Pandora's box. One dilemma was associated with the industrial investment lobby, which wished, with some justification, to measure development in terms of continued capital investment and its associated economic returns. It argued that sustained growth in conventional economic indicators (notably Gross Domestic Product) was evidence of sustainable environmental management. Indeed, there is a degree of plausibility in this argument, as continued enhancement of material prosperity would be impossible if the natural resource base did not remain reasonably intact. Defensive arguments of this nature have repeatedly enabled governments to sideline the environment as an issue whenever it was convenient to do so and to pursue a business-as-usual path of economic management. Characteristically, it has been associated with 'technical fix' approaches, in which localised environmental problems have been temporarily resolved by technological improvements (for example, in pollution control) without really addressing resource usage at a fundamental level.

The 'business-as-usual' approach has been successfully challenged latterly by persuasive scientific evidence, which has for many years been difficult to collect and interpret because of its indeterminate and complex nature. This evidence has related mainly to changes in climate, biodiversity, and condition of the global commons (e.g. atmosphere and oceans). There is a growing acceptance that, although the links between national economies and global environments are diffuse and are not reflected simply and directly through conventional economic indicators, many of our development strategies are inherently unsustainable.

A further source of difficulty associated with the WCS was its relative neglect of people. Despite an eloquent résumé of the impacts of human activities on the earth's 'life support systems' (air, soil, oceans), the original strategy failed even to include a chapter on population. Although sensitive to the plight of the world's poor, the WCS was essentially a scientific rather than a sociological analysis and it gave little recognition to the inequalities associated with trade, wealth and gender. Consequently, before the concept of sustainable development could gain the widespread endorsement of the poorer nations, it required substantial elaboration. This was first achieved by the Brandt Commission (Independent Commission on International Development Issues, 1980), whose report *North-South* advocated massive transfers of wealth to stimulate less developed economies. This attitude largely reflected a spirit of enlightened self-interest, aiming to spawn more viable trading partners. Subsequent extensions were effected by the Brundtland Report (World Commission on Environment and Development, 1987) and the second World Conservation Strategy document, *Caring for the Earth* (International Union for the Conservation of Nature and Natural Resources, 1991), which set sustainable development principles within a more balanced human and scientific context.

Sustainability thus has multiple dimensions – it is about much more than tonnes of fish to be caught or timber to be felled, or even about the physical complexities of climate change. It embraces many physical, biological, social and economic elements which regulate the interactions between human systems and ecosystems. Very broadly, environmental quality is characterised by the *conditions* or current status of these elements, the *pressures* which act upon them, and the natural or policy-driven *responses* to those external pressures. These characteristics manifest themselves in various ways, notably the

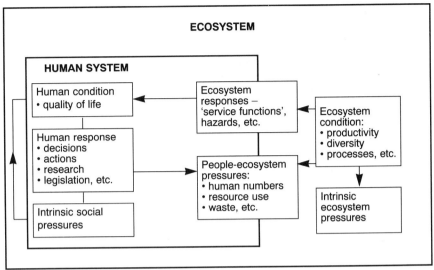

Figure 1.1   Relationships between people and the earth

Source: based on Carew-Reid et al, 1994

range of conditions which may prevail in natural and human ecosystems, the pressures exerted on ecosystems by humans, and the responses of human systems and ecosystems to external pressures. Layered on these are the intrinsic dynamic changes within ecosystems (i.e. those natural perturbations which would happen in the absence of human interference), and the pressures between groups of people which can trigger conflict or engender cooperation. These various relationships are summarised in Figure 1.1.

The environmental agenda

The environmental agenda is in practice almost limitless and includes, at least tangentially, most of the major issues currently facing humanity. The adoption of the notion of 'sustainable development' has in particular broadened the basis of concern by extending it to include issues such as war, poverty, gender, education, animal welfare, trade and health as well as purely ecological issues. However, in order to take a manageable perspective on the subject, it is useful to identify the core issues which local sustainability programmes commonly address.

One of sustainability's underlying concepts is that of a system's 'carrying capacity'. In very general terms, this implies that there is an upper limit (e.g. of human population, agricultural exploitation, recreational usage, motor vehicles) that can be supported before a system starts to deteriorate, perhaps irreversibly. These 'systems' relate to the earth's principal functions as 'source' (i.e. supplier of natural resources), 'sink' (assimilator of wastes) and 'service' (atmospheric, hydrological and other processes which provide our essential life-support). It has been proposed that the main objective of environmental management is to maintain the environment's carrying capacity on behalf of sustainable development (Holliday, 1993, quoting the first Netherlands National Environmental Policy Plan, 1989). If this notional upper capacity is transgressed, there may be irreversible effects within a generation. These may include increased mortality and morbidity amongst people, severe nuisance and damage to their well-being, the extinction of plant or animal species (reflecting a worrying underlying trend in the erosion of biodiversity) and damage to water or soil resources.

This catalogue of core environmental issues has been widely rehearsed (for a recent review, see Pickering and Owen, 1994). One important exposition, inasmuch as it relates to national policy, is in *Sustainable Development: The UK Strategy* (HMG, 1994a): more will be said about this in due course. It is not comprehensive, as its primary concern is to improve specific performance indicators in the context of an advanced post-industrial economy, but it provides a useful synopsis of key issues nonetheless. Its main focuses are very briefly summarised in Table 1.1. The WCS Strategy notes that major pressures on land arise from population growth, household formation and transport. At the same time, the economy is changing and, whilst new technologies may reduce environmental damage, pressures associated with 'cleaner' economic sectors, such as the growth in leisure and transport, might intensify. Solid waste disposal continues to be a major issue, with some 500 Mt of wastes generated annually in the UK, and we are still a long way off being the 're-cycling society' that should be the hallmark of a truly sustainable economy. Our use of energy is also closely considered, especially as energy consumption can be a revealing unit of account for the efficiency of human ecosystems.

Neither the list of environmental problems nor the government's claims of progress or future intentions should be taken at face value. Certainly the former are not of equal degrees of seriousness, nor the latter of uniform effectiveness. Controversy creeps in everywhere when discussing the environmental agenda. Long-term trends, and even day-to-day changes, present major problems of monitoring and measurement. The magnitude of change is difficult to establish and its significance – especially in relation to the earth's resilience to recover from apparent disturbance – even more so.

Despite some local environmental disasters, global catastrophe has so far failed to materialise. The extent to which environmental changes, such as ozone thinning or temperature rises, are attributable to human activities is disputed. The ability of managerial responses to curb wasteful resource use, such as recycling programmes, is also debatable. And, above all, the scale of disturbance which could occur quite spontaneously', as part of the process of natural environmental change, potentially far surpasses anything that human impacts are able to induce. This, however, is not a cause for complacency or despair, merely a warning not to misrepresent the environmental agenda, for fear of public reaction against alarmism.

Two reasons in particular should affirm the pursuit of our quest for sustainable development. One is that some environmental systems have a tendency to display sudden and (virtually) irreversible change, without giving many

Table 1.1 Summary of pressures on the UK environment

Theme	Key issues
Population growth	• modest but sustained growth until c.2030 • regional pressures through internal migration • rapid increase in number of households • need for careful use of land, include re-use of derelict land
Global atmosphere	• emissions of CO_2 and other gases could lead to climate change and sea level rises • need to review current measures for reducing 'greenhouse gases' • depletion of ozone layer by certain chemicals, leading to reduced protection from harmful solar radiation
Air quality	• reduced industrial and domestic emissions but serious pollution from vehicles • continuing problem of 'acid rain' but some evidence that SO_2-related problems becoming less serious
Freshwater	• considerable pressure on water resources in most populous regions • need for demand management in water consumption • broadly satisfactory position in relation to drinking water quality and trends in inland waters, but still serious pollution 'hotspots' • pollution from agricultural chemicals a problem in localised areas – difficult to address because diffuse nature of seepages
Sea	• identifies few problems with general marine water quality and impacts of marine dumping, though notes problems of estuaries adjoining industrial areas • some concerns about marine dredged aggregates, depletion of fish stocks and impacts of leisure activities
Soil	• notes some localised problems of soil contamination and erosion • concerns about contaminated land and its reuse
Land use	• notes that urbanised land may exceed 15% of total available in some English regions • believes planning system generally striking right balance between positive use and conservation of land • continuing concern about demands for housing, roads, etc. on 'sensitive' land
Minerals	• notes increasing difficulty of identifying extraction sites which are environmentally acceptable • some prospects for recycling various types of mineral
Wildlife and habitats	• recognises significance of UK flora and fauna associated with temperate island location and variety of habitat types • notes loss of semi-natural habitat and diminution of biodiversity
Transport	• traffic (passenger and freight) increased about threefold over last 40 years, mostly by car and lorry

	• creates problems of congestion, air pollution, noise, disturbance to communities, demand for 'sensitive' land
	• need for demand management
Energy	• need to improve efficiency of energy production and distribution
	• recognises need for careful management of wastes, including nuclear
Waste	• large quantities of wastes arising from modern production, packaging and consumption
	• threats to soil, air and water quality if wastes not properly managed
	• need to promote reuse, recycling and recovery

Source: based on information in HMG, 1994a

warning indicators. The same could prove to be true of the earth's life support systems, and we should therefore err on the side of caution in our use of them. The other reason is the continued growth of the world's population. This, coupled with the progressive industrialisation of many emerging economies, will mean that the stresses which the global environment will have to endure in the early decades of the twenty-first century may well exceed the aggregate stress placed upon it since the dawn of civilisation. Seen in this light, wantonly unsustainable development is not likely to be an acceptable option.

Sustainability, sustainable growth and sustainable development

For most of this book, the terms *sustainability* and *sustainable development* are used almost interchangeably. *Sustainability* is generally used here as a convenient shorthand for various related terms, the conventional distinctions between which may, when viewed from a practical point of view, seem precious. However, it is important to be familiar with the debate associated with the use of environmentalist terminology, and to appreciate the particular role which economists have had in developing the concept of sustainable development.

Following the WCS, a myriad definitions of sustainability were proposed (e.g. Pezzey, 1989), and alternative terms of sustainability, sustainable growth and sustainable development emerged. Sustainable *growth* was generally used to denote continued increases in GDP, subject to environmental controls. Attempts were also made to devise national aggregate indices which incorporated environmental parameters into measures of economic performance. This notion was felt by some to be a contradiction in terms as it implied that something finite (namely, the inherent capacity of the earth's resources) could grow indefinitely. Sustainable *use* has come to mean the harvesting of biologically renewable resources (crops, timber, fisheries, etc.) at a rate less than or equal to their replenishment rate. This concept is far more difficult to apply to non-renewable resources (fossil fuels, minerals). For the latter, it is broadly assumed that they are to be used on a 'lean' basis, to support human well-being, and at a rate which will continue to supply all foreseeable future needs whilst research continues into their eventual substitution. Both types of resource should be used in such a way that their environmental effects are fully accounted for.

Sustainable *development* has, after much elaboration, emerged as the most durable and central concept and is discussed more fully in the next section. 'Sustainability' has come to be used as a convenient short-hand for the spectrum of principles underlying a number of alternative terms. It also invites a convenient play on words: that the critical requirement is to build capacity in the *ability* of people and their representative organisations to *sustain* lifestyles which are compatible with continued environmental integrity.

Basic principles of sustainable development

'Sustainable development', has emerged as the most defensible and elegant term. Economists, in particular, have resolved much of the controversy and confusion which has clouded our discussions of its meaning and application. Sustainable development is now conventionally associated with three fundamental principles.

First, the principle of 'inter-generational equity' (futurity) implies that one generation should hand on the earth to the next generation in at least as good a condition as it inherited it. This is merely a technical expression of a long-standing sentiment. It is neatly summed up in the proverb that 'we have not inherited the earth from our parents, we have borrowed it from our children'. It is also reminiscent of the archaic legal term 'usufruct', which was used in this context by the seventeenth-century diarist John Evelyn, requiring that a borrower of an item returns it to the lender in its original condition.

Second, the concept of 'intra-generational equity', or social justice, requires that sustainable development contains within it a principle of human needs. The Brundtland Report, for example, had particular regard to the needs of the world's poor, to whom it was argued that an 'overriding priority' should be given.

Finally, the principle of 'transfrontier responsibility' states that sustainability in one locality, region or country cannot be achieved at the expense of environmental conditions elsewhere. Thus, we must accept responsibility for any impacts that our activities may have on the water and air quality, biodiversity and the condition of the natural resource stock in other areas. Often, these areas may be very distant, and their environmental deterioration may be unrecognised.

These principles are succinctly encapsulated in the most widely adopted definition of sustainable development (the Brundtland definition):

> development that meets the needs of the present without compromising the ability of future generations to meet their own needs' (WCED, 1987).

Even this elegant dictum is divisive: some see it as too anthropocentric, whilst others admire its essentially human focus. Their dissension pivots upon whether we consider humans to be an inherently higher life-form and thus, at least metaphysically, separate from nature. It is an absorbing debate, which draws heavily on ethical and religious concepts, but it cannot be explored here. At the risk of alienating 'deep greens', the Brundtland definition is broadly accepted as a working hypothesis.

The Brundtland Report coincided with a rapid increase in the level of serious international attention to environmental matters, largely precipitated by accumulating scientific evidence about human-induced change in critical natural systems. Considerable effort was duly focused on converting the concept of sustainable development into a practical basis for decision-making. The most rapid advances on this were made by neo-classical economists, reflecting a broad political mood (reinforced by the contemporaneous collapse of East European communism) to work with markets rather than purely through regulation. Thus, whilst environmental management has traditionally relied on regulatory (command and control) measures, the more recent trend has been to use market mechanisms to promote sound resource use.

Admittedly, in recent years regulation theorists have sought to dispel simplistic faith in market-based solutions by demonstrating the ways in which the stability of capitalist societies is reliant on complex and irreducible regulatory mechanisms. The two approaches can clearly be used in a complementary fashion in the pursuit of sustainable development. Both are now strongly influenced by the conceptual frameworks proposed by economists, with economic concepts often being used to reinterpret the justification for various modes of state regulation.

Economists view the Earth's resources in terms of capital, which can be conveyed as bequests between generations, and they make an important distinction between *constant* (or substitutable) natural capital, and *critical* natural capital. The former refers to cultural (e.g. built), and readily renewable natural, resources. The latter includes basic life-support systems (global commons, biodiversity) which, once irreversibly damaged, cannot be re-created. Collectively, the two represent the Earth's natural capital stock.

Some would argue that, provided the net quantity of the stock is conveyed from one generation to the next, the conditions of sustainable development are satisfied. However, this is described as 'weak sustainability' as it is indifferent to the form in which the capital stock is transferred, and has no regard for the balance between critical and constant stocks. For example, an ancient wetland may be destroyed in the course of development and be substituted by an equivalent area of newly created water space and fringing vegetation. This might appear similar and provide an equivalent level of amenity, but may entail a serious, hidden loss of biodiversity, especially at the unseen scale of micro-organisms. Valuable and essentially irreplaceable resources of this kind comprise critical natural capital, as they cannot be replicated during the period of one, or even several, generations. Thus, a 'strong sustainability' stance requires not only that the total capital stock is at least as generously bequeathed as received, but also that, within it, the component of critical natural capital remains substantially intact.

Three factors dictate that we take the strong sustainability rule seriously (Pearce, 1993). First, there is uncertainty about the highly complex way in which environmental systems work and thus we must be cautious about any actions which might cause them to be disrupted. This requires that we adopt a 'precautionary principle' towards our use of critical natural capital. Second, changes in the earth's life support systems may prove to be irreversible. There may, for instance, be a point at which rises in global temperature take on a momentum of their own and cannot be forestalled, or at which entire seas are unable to recover their cleansing and regenerative capacities. Third, some components of the environment are essential to human well-being even if not to survival. Humans need a degree of naturalness and complexity in their surroundings: thus, loss of assets such as amenity and wilderness may be as important as more fundamental biological ones.

In principle, there is widespread agreement with this analysis. However, when presented with actual decisions which involve the incremental loss of critical natural capital, a willingness to make the appropriate sacrifices may be less forthcoming. Even the most environmentally-conscious governments will wish to continue to deliver an acceptable quality of life (not least in material terms) to its citizens. This potential contradiction between natural capital conservation and economic growth is addressed by Lowe and Murdoch (Rural Development Commission, 1993) who argue that achieving growth itself is not the problem, but rather 'environmentally unconstrained growth'. A more awkward question is that of the social priorities of economic development. Thus, whilst sustainability implies an unarguable need to integrate environmental considerations and economic policy-making, it is apparent that economic growth for some may be achieved at the expense of the economic well-being of others. Associated with each of our previous principles of futurity, intra-generational bequests and transfrontier reponsibilities, therefore, will be distributional consequences, benefiting some at the expense of others. It is clear that sustainable development is about much more than the environment, involving consideration of fundamental questions of justice and democracy as it does. This, far from being an insurmountable complication, is one of its great strengths, and one of the reasons why sustainable development will not evaporate as a guiding framework for communities and governments.

Without, at this stage, entering into these deeper issues, the UK Government (HMG, 1994a) claims that the dilemma is broadly soluble. The reasons for this assurance may be summarised in the following terms:

- Sustainable development does not necessarily mean having less economic development.
- A healthy economy is better able to generate the resources to meet people's needs.

- New investment and environmental improvement often go hand in hand.
- It does not mean that every aspect of the present environment should be preserved at all costs.
- It does mean that decisions throughout society are taken with proper regard to their environmental impact.

The limitations of these principles in terms of their anthropocentric, western-ised perspective need not detain us for the present. They do serve to demon-strate, in the words of the same document, that the purpose of sustainable development is one of reconciling two basic objectives:

- Most societies want to achieve economic development to secure higher standards of living, now and for future generations.
- They also seek to protect and enhance their environment, now and for their children.

These tenets have a ring of indisputable authenticity but, as we have already obser-ved, they may be put to their sternest test when specific disputes arise at a local level.

After three decades of dispute about the need for effective environmental management, a consensual position is thus being reached. In broad terms, it appears that sustainability will entail a popular commitment to couple social and economic welfare to environmental capacity. This has some specific im-plications at the local level. It will require that we develop comprehensible principles for practical everyday action, that we achieve attitudinal change in respect of individual behaviour and lifestyle, and that we cultivate effective means whereby citizens and communities can participate in critical decisions.

Practical guidelines for local sustainability

Whilst the need to pursue sustainable development seems almost disconcer-tingly overwhelming, we must still be careful to examine critically the basis of this quest. We have already noted the controversial nature of the scientific evidence and the disputed basis of public policy. If we are uncritical in our pursuit of sustainability we may end up being guilty of foisting our own views on other legitimate interest groups (environmental imperialism) or of suppress-ing reasonable human rights and expectations in the pursuit of environmental goals (ecofascism). Just and workable solutions must be sensitive to the inter-ests of a spectrum of user groups.

A fair and defensible set of requirements to implement sustainable develop-ment is difficult to define, but some reasonably constant principles are now available. The following legal guidelines have been proposed (Carew-Reid et al, 1994), with the advice that they are best considered together, as the rigorous application of one principle in isolation might be impracticable or inequitable. In summary, these guidelines comprise:

- The *public trust doctrine*, which places a duty on the state to hold environ-mental resources in trust for the benefit of the public.

- The *precautionary principle* (erring on the side of caution), which holds that where there are threats of serious or irreversible damage, lack of full scientific certainty shall not be used as a reason for postponing cost-effective measures to prevent environmental degradation.
- The *principle of inter-generational equity* which requires that the needs of the present are met without compromising the ability of future generations to meet their own needs.
- The *principle of intra-generational equity*, stating that all people currently alive have an equal right to benefit from the use of resources, both within and between countries.
- The *subsidiarity principle*, which deems that decisions should be made by the communities affected or, on their behalf, by the authorities closest to them (though the 'appropriate level' of decision-making may be problematic to determine in respect of international environmental issues or transboundary pollution).
- The *polluter pays principle*, which requires that the costs of environmental damage should be borne by those who cause them; this may include consideration of the damage occurring at each stage of the lifecycle of a project or product.

The foregoing are fundamental and elegant principles, which should direct and influence the making of laws and policies. On an operational basis, more pragmatic and routine measures may be necessary. For example, the Countryside Commission (1993) has argued, more simply, that:

- policies should be consistent with environmental needs;
- each activity should contribute to the achievement of national and local environmental goals;
- activities should be carried out so that they do not damage important environmental features or exceed the environmental capacity of an area or resource system; and that
- irreversible environmental damage should be avoided where the consequences of action are uncertain and the impact is likely to be substantial.

A further set of working principles (distilled from HMG, 1994a and Blowers, 1993b) is summarised in Table 1.2

A particular problem is posed by non-renewable natural resources such as minerals and fossil fuel, which are essential to economic development and yet whose use is inherently unsustainable. Various ground rules have had to be drawn up to establish working principles for sustainability in this context. These principles seek:

- to conserve non-renewable resources as far as possible, whilst ensuring an adequate supply to meet the needs of society;
- to minimise the production of wastes and to encourage efficient use of non-renewable resources, including carefully targeted use of high quality resources, and maximum use of recycled wastes;

- to encourage sensitive working practices during and after the extraction of non-renewable resources; and
- to protect areas of critical landscape or wildlife value other than in exceptional (and precisely defined) circumstances.

(Adapted from DoE, 1996).

In broad terms, this implies minimising the irretrievable and irreversible loss of key resources, so that their extraction, processing and consumption are limited to what is absolutely necessary to meet the needs of the current generation. This aim is assisted by improved understanding about, and availability of, the use of recycled materials in construction work and other areas of economic activity. Moreover, the overall quality of the environment affected by the extraction of non-renewable resources needs to be conserved or improved over time, so that future generations are not disadvantaged by the activities of the present one.

Table 1.2   Some working principles for sustainability policy

- decisions should be based on the best possible scientific information and analysis of risks
- where there is uncertainty and potentially serious risks exist, precautionary action may be necessary
- ecological impacts must be considered, particularly where resources are non-renewable or effects may be irreversible
- cost implications should be brought home directly to polluters or waste producers
- we should seek to conserve the stock of natural assets, wherever possible offsetting any non-essential reduction by a compensating increase
- development should avoid damaging the regenerative capacity of the world's natural life-support systems
- sustainable development must aim to achieve greater social equity
- we must avoid imposing added costs or risks on succeeding generations.

Source: based on information in Blowers, 1993b and HMG, 1994a

The various principles of sustainable development must be worked into specific instruments which give effect to general policy. These vary in relation to the extent to which they apply to national and local levels, but the main types of approach (see Table 1.3) essentially seek to limit the magnitude of environmental impacts by control and influence. Broadly speaking, historical practice has been to seek to satisfy demand and thus increase the supply of goods and services. Sustainable development requires that levels of demand must be at least partly restrained. Thus, there may be explicit measures of 'demand management' – some based on fiscal or economic means, and essentially the purview of national governments, and others, more locally inspired, based on physical measures. Many of the other instruments – controls, education and information provision, and promotion of more responsible stewardship – are applicable to both national and local tiers, but are often most effectively delivered locally.

Table 1.3 Main instruments for sustainable development policy and practice

Type of legal/policy instrument	Examples of approach
Demand management	• development plans and other land allocation mechanisms • economic incentives, charges and taxes to reduce resource use (e.g. incentives for recycling, levying supplementary charges on road users in congested urban areas) • regulatory powers (e.g. enforcement of fixed pollution standards) • physical management techniques (e.g. traffic calming)
Control of land use change	• environmental assessment of proposed new projects
Environmental information provision	• corporate environmental audits state of environment reports
Internalisation of environmental costs	• pricing mechanisms to mitigate damage or compensate sufferers (i.e. polluter/user pays) through e.g. carbon taxes
Primary environmental care	• direct community action changing individual behaviour

Summary

Following this introductory review of the dimensions of sustainability, the next chapter outlines the broader framework within which local sustainability is set. Thus, the role of the locality is set within the context of global, continental and national agendas for environmental management. Earlier, we noted that localities could not simply adopt a veneer of sustainability, by exporting their demands elsewhere. Thus, in Chapter 3, we examine the nature of sustainable 'places' – regions, cities and rural areas – and consider the factors which influence the scale and intensity of their environmental impacts. A great deal of research and practice has taken place recently on ways of monitoring and managing sustainable development. Chapter 4 takes up this theme by addressing the methods and techniques associated with environmental reporting and decision-making. In addition to the formal methods associated with 'rational' management, it also considers some of the more qualitative and socially based approaches suitable for local application.

The second half of the book turns to the roles of the key players in local sustainability. The role of local government is pivotal, at least in instigating sustainable development initiatives. Chapter 5 therefore addresses both the role of local authorities generally and the approaches which they are adopting to 'green' their own activities. Although a wide range of local government responsibilities is relevant to sustainable development, the town planning profession has a unique role in this regard, both in terms of its statutory remit and its broader philosophy. The contribution, both potential and actual, of the planning system is examined in Chapter 6. This is followed by an account of the business sector which drives the local economy and without whose conversion to 'green' behaviour all programmes for local sustainability will founder. Finally, the role of the individual citizen is addressed. Environmental ethics and behavioural principles have transformed our understanding of responsible citizenship and an exploration of the rights and duties of the individual forms a fitting conclusion.

2

The context of local sustainability

From village greens to global commons

One of the most compelling critiques of the unsustainability of contemporary society was Garrett Hardin's essay on the *tragedy of the commons* (Hardin, 1968). From time immemorial, in most countries, some areas of land, water and forest have been held on a communal basis. In these areas, particular groups of people are permitted to exercise specified user rights, for example to graze domestic livestock, fish and gather fuel. Access to a common resource is often regulated by a representative group of commoners.

The essence of Hardin's thesis was that, over time, commons would become over-exploited. There is presumably an upper limit of utilisation to which a common can be subjected and, within this overall demand, each user can take a fair share of a constantly renewable resource. There are always likely to be a few selfish users who will take more than their entitlement, but this small level of over-exploitation is unlikely to lead to irreversible deterioration of the common. However, if many commoners follow suit, which would be a natural human response, the resource will become seriously over-exploited. This gloomy prognosis suggests that all common resources will, eventually, be characterised by a scramble for usage.

This may seem a minor matter, but its basic principle can be extended much more widely. In particular, the analogy can be extended to so-called 'global' commons, such as the air, oceans and fisheries. Throughout much of the world, these presently appear to be suffering from selfish exploitation, short-term gain and collective misuse.

There are a number of criticisms of this hypothesis. On the one hand, there is much evidence of very careful regulation of access to and use of some commons over long periods of time. Equally, over-exploitation is often caused through neglect, poverty or ignorance, rather than greed. Also, the analogy between local commons and 'global' commons is often imperfect (except, perhaps, with the atmosphere), as the latter are often affected by access restrictions and ownership rights. Nevertheless, the general principle is helpful in understanding some of the problems facing sustainable development.

A basic feature of common resources is that commoners are not charged for their use. A characteristic of much of the earth's 'critical natural capital' is that it is unpriced or seriously underpriced. This relative freedom of access tends to encourage unsustainable use. In broad terms, therefore, the 'tragedy of the commons' provides a useful insight into our utilisation of critical resources.

Hardin's thesis was principally directed at life-sustaining productive resources and environmental systems. Yet even in the case of amenity and aesthetic resources, there may be a comparable argument. Cultural landscapes – those landscapes which, in their scale and appearance, appear to have been 'built by hand', and which are locally distinctive – convey, between generations, both a valued human asset and (fortuitously) an exceptional bequest of biodiversity. They also betray evidence of 'common' land management practices, which have often been fine-tuned over the centuries in response to local conditions. These areas of land, too, are seriously undervalued in market terms and are liable to thoughtless destruction. As well as producing a cultural heritage of fine quality, which is effectively irreplaceable, these traditional practices may furnish exemplars for living sustainably within fixed environmental capacities. When these practices become economically obsolescent, however, and people abandon them, serious difficulties arise in sustaining the valued landscapes which depended on their labour intensive management regimes.

During the 1970s and 1980s, environmental literature was dominated by accounts of international and global issues. This was a necessary emphasis if politicians and transnational corporations were to be convinced of the scale and seriousness of current environmental damage. However, its corollary was that the local dimension was relatively neglected. There was a real risk that individuals, who might be deeply concerned about environmental issues, would nevertheless tend to experience a sense of remoteness from the power-brokers of change, and of futility in making a personal response. It is easy for individuals to identify with the loss of village greens and local commons (and, by analogy, other locally valued assets), but much more difficult for them to comprehend or feel a sense of urgency about the thinning of parts of the stratospheric ozone layer. Whilst it is clearly essential to proceed with international treaties and protocols, pursuit of these alone means that responses by individuals and communities to make their own lifestyles more sustainable may be reduced to no more than the level of a gallant and altruistic protest.

There have, of course, been many examples of grass roots environmental action over the past three decades. These have ranged from official attempts to engage local communities in politically acceptable practical tasks, through to (politically unwelcome) protest against the state and corporate enterprise. Leading conservation agencies were quick to popularise the slogan 'think globally, act locally' as a way connecting personal and community actions to larger scale endeavours. However, slogans are not always accompanied by the explicit mechanisms necessary to put them into effect.

Whilst governments have always valued the spadework put in by local communities, they have generally been resistant to the idea that effective responsibility or power should be transferred down the hierarchy. Indeed, recent years have seen a significant centralisation of power in many (though not all) countries, and an erosion of the status of local jurisdictions, such as districts, parishes and communes. The reasons for this have included the pursuit of tighter fiscal controls, a belief that effective action is only possible through strong decision-making at the centre, and perhaps a mistrust of the capability of local bodies to take dispassionate long-term standpoints on key issues. By contrast, we have previously referred to the principle of 'subsidiarity' (see page 15), which requires the maximum possible transference of power down the bureaucratic hierarchy, as one of the pre-requisites for sustainable decision-making.

Authoritative analyses of sustainable development suggest that as much as sixty per cent of the content of action programmes will need to be addressed at a local level. Whilst much of this action will lie in the domain of local government, this should not imply that local initiatives should be reliant on official bodies, although democratically accountable bodies must play a key role. It is clear that all sustainability initiatives must win the hearts and minds of individuals, and not just the support of officialdom, if they are to work in practice. In developed countries, local debate and action may be necessary to achieve quite profound attitudinal change, as citizens may need to accept a perceived deterioration in some material expectations and this change of attitude is most unlikely to occur if the exhortation comes in a high-handed manner from government. In developing countries, sustainability may involve investment in programmes, and the adoption of technologies, which are sensitive to particular circumstances and may be derived from indigenous sources and local knowledge.

Thus the local level – and its governmental bodies, workplaces, interest groups and individual citizens – is crucial to the attainment of sustainability. Perhaps some of the lessons of well-managed commons can assist us in this respect. Common resources have been passed between generations in good condition where there is a general respect for their value and effective regulation of use. O'Riordan (1995) has also drawn attention to the ways in which users of common resources typically display high levels of mutual support. It would be wrong to advocate too much of a 'parish pump' mentality, to the exclusion of broader issues. Indeed, local citizens must have an interest in, and even passion for, the 'global village'. Nevertheless, a characteristic of success will be the extent to which individual people and their neighbourhoods define and pursue their own paths towards more sustainable lifestyles. This must occur within the context of 'thinking globally, acting locally', and so we now turn to the concentric shells within which local action is set.

The global context

The interconnectedness of environmental systems requires concerted international responses to key issues. International government is, however, poorly developed, especially on topics other than aid and trade. Consequently, there has been a series of conferences, treaties, conventions and protocols to try and address the more urgent 'transboundary' environmental issues. Of particular interest to sustainability have been environment and development strategies aimed at effecting general and specific responses from national governments. These have often been launched by high profile conferences.

The global dimension has been addressed in considerable detail by many texts, and it is only summarised here in outline. Very broadly, global initiatives fall into 'multi-sectoral' programmes, treating the environment in a more or less holistic fashion, and 'sectoral' or 'thematic' strategies, which address single issues (Carew-Reid et al, 1994). Whilst they tend not to be associated with specific legal and funding measures in their own right, they are frequently instrumental in triggering additional controls and incentives. A recent key example of this has been the creation of a Commission on Sustainable Development (CSD), recommended by the UN Conference on Environment and Development, to monitor progress on Agenda 21. The key international strategies are summarised in Table 2.1. Although not mentioned, the work of the Paris-based Organisation for Economic Co-operation and Development (OECD) should also be recognised for its increasing efforts to integrate environmental considerations into economic development issues.

Table 2.1   Some international strategies for the environment

Multi-issue	Single issue
Stockholm Conference Action Plan (UN, 1972)	Global Biodiversity Strategy (WRI/UNDP/UNEP, 1992)
World Conservation Strategy (IUCN/UNEP/ WWF, 1980)	Tropical Forestry Action Programme (FAO/ WRI/WORLD BANK/UNDP, 1987)
Report of World Commission on Environment and Development (Our Common Future/ 'Brundtland' Report) (WCED, 1987)	Strategy and Agenda for Action for Sustainable Agriculture and Rural Development (FAO, 1991)
Report of Latin American and Caribbean Commission on Development and Environment (Our Own Agenda) (UNDP/IADB, 1990)	Global Strategy for Health for All by the Year 2000 (WHO, 1981)
	Plan of Action to Combat Desertification (UNCOD, 1977)
Caring for the Earth: a strategy for sustainable living (IUCN/UNEP/WWF, 1991)	World Population Action Plan (WPC, 1974)
Agenda 21 (UNCED, 1992)	Strategy for the Protection of the Marine Environment (IMO, 1983)
Various strategies for shared regions, such as 'regional seas programmes' and river basin strategies.	Climate Change Strategy (WMO/UNEP, 1992)

Source: based on Carew-Reid et al, 1994

The most definitive and widely influential of these statements has been Agenda 21, arising from the UN Conference on Environment and Development. This has both set the agenda across a huge spectrum of policy and management issues for the twenty-first century and has more broadly inculcated an acceptance of sustainability based on the inseparability of social, economic and environmental issues. The Conference also instigated three global statements on key environmental concerns, namely:

- the Framework Convention on Climate Change
- the Convention on Biological Diversity, and
- the statement on Principles of Forest Management.

The products of the Conference were thus lengthy and wide-ranging, though it is sufficient for the moment merely to be aware of the general nature of Agenda 21. The main report is organised into sections which address the four major areas of political action (Grubb et al, 1993), namely: social and economic development (Chapters 1–8); natural resources, fragile ecosystems, and by-products of industrial production (Chapters 9–22); major organisations and groups of people (Chapters 23–32); and means of implementation (Chapters 33–40). Each programme area is normally organised into sections reviewing the basis for action, the objectives to be addressed, the activities needing to be undertaken, and the means of implementation, both technical and financial. Both the main agenda and the three global statements have specifically been followed up by those national governments averring a comprehensive commitment to UNCED.

The European context

Environmental policy is now often driven at the continental level, and this is especially true in Europe. At a constitutional level, the European Union influences the legislative and budgetary frameworks of its member states. One of its strongest areas of legislative influence has been the environment, especially certain pollution control and land use issues. Although the UK has a long history of effective practice in all aspects of environmental management, it would have been unlikely to move towards international best practice had it not been for European pressures. Of particular significance to the present account is the EU's Fifth Environmental Action Programme *Towards Sustainability* which is subject to periodic review and updating (Commission for European Communities, 1992).

Broadly, five areas of environmental policy are addressed at European level: industrial pollution, productive land management (agriculture and forestry), nature conservation, environmental assessment and urban change. Environmental issues may also be influenced by other policy areas, such as regional development. Although the breadth and generality of European policy statements and legal instruments may seem remote from the local level, the EU has been an enthusiastic exponent of subsidiarity and expects that implementation will proceed at national and sub-national levels. Official EU policy is shaped by periodic (roughly quinquennial) Environmental Action Programmes (EAPs). These have gradually shifted from reactive (end-of-pipe) approaches to pro-active (anticipatory) ones. All have relied on the imposition of fixed standards to drive improvements in the quality of air and water; the traditionally more flexible British approach has gradually converged with the continental ap-proach in this respect. Some of the key continental influences on sustainability are summarised in Table 2.2.

Table 2.2   Key elements in the European 'sustainability' framework

Document	Major concerns
EAPs 1–3	basic aims and principles for pollution control and clean-up actions; gradually taking a longer-term view and seeking overall improvement in environmental quality
EAP 4	environment raised to issue of central importance; becomes a key component in economic, industrial, agricultural and social policies
EAP 5 (1993–2000)	stresses need for more proactive approach in order to alter behavioural patterns and to move towards sustainable development; continues with centralised regulatory approach; makes more money available for environmental projects; sees sustainable development as the basis for all future developments in the EU
Agriculture	'greening' of policies: environmentally sensitive areas, extensification, nitrate controls
Forestry	Helsinki conference agrees to forest objectives for: conservation and appropriate enhancement of biodiversity; management aimed at increasing the diversity of forest habitats; favouring native species
Environmental impact assessment	standardised approach to assessment of the environmental effects of major development projects throughout EU introduced by 1985 directive; directive on environmental assessment of projects, policies and plans ('strategic' environmental assessment) to follow
Europe 2000	EU 'Green Paper' encouraging more sustainable cities; assumes compact settlement patterns to be most efficient

Source: partly based on Gibbs, 1993

The national context

It is the national context – of legislation, policy and government structures – which really frames the scope for local action. The UN Conference on Environment and Development recognised nation states as the first point of cascade for Agenda 21, and thus urged that:

> . . . each country should aim to complete . . . if possible by 1994, a review of capacity, and capacity-building requirements, for devising national sustainable development strategies, including those for generating and implementing its own Agenda 21 action programme'. (UNCED, 1992)

National documents were generically referred to as 'national sustainable development strategies', although more euphonious titles have been adopted for actual documents. In addition, each signatory to the agreements on climate change, biodiversity and forestry was expected to produce national policy statements on those subjects.

In the UK, a suite of documents was produced to follow up UNCED, namely (HMG, 1994a–d):

- Sustainable Development: the UK Strategy (the National Sustainable Development Strategy).
- Biodiversity: the UK Action Plan.
- Climate Change: the UK Strategy.
- Sustainable Forestry: the UK Strategy.

These were received with varying degrees of acclaim by the environmental lobby. Although all were cautiously welcomed as wide-ranging and progressive statements, they did not always set the radical agenda that many had desired. A common feature, especially associated with the Climate Change document, was that they tended to summarise and repeat existing policies rather than generate new ones. Some critics compared them unfavourably with other national approaches. A widely acclaimed national strategy, pre-dating the Agenda 21 process, was the Dutch National Environmental Policy Plan (Tuininga, 1994; VROM, 1993), which set tough targets against tight time-scales for a variety of improvements to pollution and resource usage. It also aimed to engage various 'target groups' (transport, energy and industry sectors) in a dialogue with government, and this has appeared to be highly productive. Although there are implementational problems associated with this plan, some important progress is being made in the management of industrial, trade and agricultural activities, and in integrated life-cycle management.

Despite its flaws, the UK Sustainable Development Strategy provides a comprehensive baseline of environmental conditions and policies, and a framework into which local initiatives can be placed. It also defines the role of central government in effecting the Strategy, which is particularly significant in an age when we recognise that mere decrees from government cannot induce fundamental change in the economy or in the ways people behave. The role of central government thus lies much more in terms of setting environmental quality targets, as these relate to different media of the environment (e.g. air, freshwater) and different sectors of the economy. These targets become more meaningful in the context of environmental indicators (i.e. summary indexes of environmental quality and policy performance) which government, in consultation with interested parties, can also set. The government can also introduce and refine regulatory mechanisms such as pollution and planning controls, and market-based economic instruments which influence the behaviour of consumers and suppliers.

However, the UKSDS is quick to affirm that the role of central government is limited, and that the emphasis must be on collaborative working. Key contributors to the national debate on sustainability are the Government's Panel on Sustainable Development and the UK Round Table on Sustainable Development. Citizens' views and activities are increasingly to be incorporated through the 'Going for Green' programme, launched in 1995. Citizens can be identified as belonging to various interest groups, which often overlap, and these comprise their roles as green consumers, householders, volunteers, workers, and parents and others involved with children. The government claims to seek the active participation of each of these groups.

The Strategy also has a more speculative, imaginative (but perhaps politically unpalatable) agenda, which may well move us along the road to deeper sustainability. This places a heavy emphasis on transport and energy usage. For instance, it considers the case for specific market measures such as congestion charging, that will increase the marginal cost of transport to the user and reflect the wider costs to others. Similarly, it suggests that the town planning system could potentially seek to ensure that land is developed in ways which reduce the need to travel and thus lead to a reduction in exhaust emissions. It notes that local authorities are also in a position to help secure integrated packages of transport and land use policies for walking, cycling and the efficient use of public transport. For such radical measures as these to be implemented at the local level there would need to be clear financial and legislative support from central government. They would also need to be accompanied by a substantial raising of public awareness about the implications and environmental costs associated with the growth in transport emissions, and of the need to modify our behaviour.

The Biodiversity Action Plan describes the range of biological variation in Britain, and proposes an action plan and broad work programme to address a range of key issues, 'biodiversity' being defined as 'the variability among all living organisms from all sources including *inter alia* terrestrial, marine and other aquatic ecosystems and the ecological complexes of which they are part; this includes diversity within species, between species and of ecosystems'. The Action Plan was generally well received, despite some misgivings about continued damage to wildlife habitats and the willingness of Government to follow up its commitments with additional primary legislation.

Amongst the major issues which it considers is the vitality of wildlife both within specific habitats such as nature reserves, and in the wider countryside. This latter aspect is considered to be of increasing importance, as purely site-based conservation strategies have proved to be ineffective in stemming the loss of biodiversity. More emphasis needs to be given to the overall matrix of patches and corridors throughout the ordinary farmland, woodlands and open ground of the countryside. Second, a more sustainable approach is urged to the use of natural resources which contribute to biodiversity, such as woodland, coastal areas, fresh and marine waters and farmland. Third, there is perceived to be a national responsibility towards biodiversity internationally, and thus there are commitments to support conservation initiatives overseas. Other issues relate to information and awareness, particularly the facilitation of environmental education and the provision of high quality and readily accessible environmental datasets.

Conservation and enhancement of biodiversity is seen to rest on core underlying principles which relate to broader issues of responsible environmental management. These dictate that biological resources should be used sustainably, that non-renewable resources should be used wisely, that individuals and communities should be involved in caring for biodiversity, and that biodiversity conservation should be included as an integral part of government programmes, policies and plans. Three key objectives for conserving biodiversity are set out in Table 2.3.

Table 2.3 Key objectives for biodiversity conservation

1. To conserve and where practicable to enhance:
- the overall populations and natural ranges of native species and the quality and range of wildlife habitats and ecosystems;
- internationally important and threatened species, habitats and ecosystems;
- species, habitats and natural and managed ecosystems that are characteristic of local areas; and
- the biodiversity of natural and semi-natural habitats where this has been diminished over recent past decades.

2. To increase public awareness of, and involvement in, conserving biodiversity.

3. To contribute to the conservation of biodiversity on a European and global scale.

Source: HMG, 1994b

The UK Strategy on Sustainable Forestry stemmed from the accord on Forest Principles at UNCED. Whilst the UK Government was one of the countries which would have preferred a legally binding convention on forests, this was opposed by a number of leading timber producing nations, and a less cogent set of principles had to be accepted. Nevertheless, these can be translated into useful national frameworks and standards. Britain's forestry policy, having developed over some eight decades, initially with a very strong emphasis on large-scale commercial planting, has now become very definitely one of multiple use and diversification of the national forest estate. The Strategy summarises domestic policy as being concerned both with the sustainable management of existing woods and forests, and a steady expansion of tree cover to increase the many diverse benefits which forests provide. The main programme areas covered by the Strategy include:

- protecting forest resources;
- enhancing the economic value of forest resources;
- conserving and enhancing biodiversity;
- conserving and enhancing the physical environment;
- developing the opportunities for recreational enjoyment;
- conserving and enhancing our landscape and cultural heritage;
- promoting appropriate management; and
- public understanding and participation.

The contemporary forestry agenda, despite still having to meet tough financial targets, is thus strikingly different from former policy. For much of its history, forestry policy was driven purely by economic considerations, with an assumption that the forest estate would be planted mainly in the uplands and composed of a very limited variety of productive and tolerant species. Thus, projects such as the Highland Birchwood Initiative, which aims to regenerate native hardwoods, have only recently become possible. The Sustainable Forestry Strategy was generally well received, and it is a clear reflection of the broad acceptance of the desirability of investing public money in multi-purpose woodlands. These woodlands are likely to contain much higher proportions of native broadleaved species (or Scots pine in the far north), to accommodate biodiversity in forests, and to integrate a range of rural land use objectives.

The final 'sustainability' document, on Climate Change, was rather less well received. To many critics, it appeared to lack coherence and vision, and it dwelt substantially on individual 'showpiece' energy efficiency projects. The Strategy's general thrust was to advocate a precautionary approach to the management of future atmospheric quality, recognising that there is sufficient evidence of climatic threat to warrant immediate action. In common with other strategies, the document was prepared on a partnership basis. Thus, consultees included consumer groups, environment groups, housing and transport organisations, local authorities, business organisations, energy utilities and central government.

The Strategy placed great reliance on the operation of the market in making progress towards more efficient energy use and improved air quality. Relevant mechanisms included the removal of market barriers and 'sending the right price signals' to consumers. A further emphasis was on the role of the public sector in providing a lead to industry and commerce. Consequently, the document also reviewed various local authority initiatives for CO_2 reduction from their housing stock, and for traffic management and improvements in driver behaviour. It also commended the use of energy audits, and the 'Making a Corporate Commitment' campaign, in which major employers set targets for energy and transport reduction. Timely recognition was also given to the role of woodlands in acting as carbon 'sinks'; this obviously feeds back to the Forestry Strategy, and provides a further justification for public support of new afforestation.

The local context

We have already seen that Agenda 21 (in 1992) urged local governments to embark on the production of local equivalents, with the majority to be in place by 1996. The intervening four years saw a flurry of activity as many councils commenced their task. Although some had completed their local strategy document as early as 1994, the rate of readily observable progress was not as rapid as might have been anticipated. This was because councils were too familiar with the pitfalls of issuing documents for public consumption in a 'top-down' fashion, only to find that they met with indifference or worse. It was of the greatest importance to ensure that people living within the council's area felt that they 'owned' their Local Agenda 21. Thus, most exercises emphasised 'process' as much as 'product', so that the commitment of a whole range of stakeholders to a shared set of principles and actions could be secured.

More detailed coverage of the role of local government in promoting Local Agenda 21s is given in Chapter 5, and the present account is no more than a preview. An important contextual consideration is the characterisation, by some political commentators, of the 'local state' in terms of its focus on consumption rather than production issues, with the latter being the domain of central government. Thus, the local arena is frequently one in which residents voice their concerns about quality of life. These are related directly to the existence of ambient environmental amenities, and indirectly to our patterns of consumption of stocks of environmental capital. In the late twentieth-century, environmental amenity is one of the key 'commodities' which is 'consumed'. Consequently, we find that local government is an important arena in which commodity issues are debated and bargained. It has also become a key proponent of techniques for environmental planning and auditing and of 'capacity-building' methods: these latter assist local communities to participate in, or even take lead responsibility for, decisions which affect their neighbourhoods.

An important component of local government, which also receives special coverage later on, is the town planning system. This grew from nineteenth-century concerns about unhealthy and crowded industrial cities, and was very much associated in its early days with the creation of garden cities. During its growth as a major profession in the twentieth-century its concerns have broadened to include management, design, engineering and various other topics, but its focus has never drifted too far from environmental quality. Indeed, the pioneers of early town planning had strong interests in human ecology, and this forms a very natural platform for modern concepts of sustainable development.

At this stage, it is worth drawing attention to some key elements of the town-planning system relevant to local sustainability. First, there is general regulation of the development of land (not including farming or forestry, but including most other building and engineering operations), which can ensure that land is not wastefully used and that critical land resources are substantially protected. Second, it enables a long-term timescale to be adopted, which – although politically mediated – can transcend short-term political expediency in the consumption of resources. Third, it permits the inclusion of public viewpoints in the preparation of plans and in some development decisions and some planning authorities have utilised quite imaginative and radical approaches to achieve this end. Fourth, it allows for the integration of various land use interests in a 'horizontal' sense, cutting across traditional professional divisions such as engineering, ecology and economic development. Fifth, since 1991, plans have been required to consider the physical environment (quite broadly understood) and, indeed, may now be 'environment-led'. One final point is the geographically comprehensive nature of planning, particularly since the 1991 Planning and Compensation Act, which required complete territorial coverage of local plans. Overall, governments have vaunted the planning system as a major element in the national pursuit of sustainable development and, in England and Wales, through its use of Planning Policy Guidance Notes, the Government has explicitly sought to include many aspects of sustainability in local authority planning practice.

Although the town-planning system has a special place in local sustainability, it is not unique. Several other areas of responsibility of local government have direct relevance, particularly environmental health, transport, leisure, and education. The collective contribution of local government to sustainability must therefore be approached on a corporate basis. Consequently, many local authorities have appointed environmental officers, often placed within the Chief Executive's Department and thus able to exercise an authority-wide brief.

Local government has also become a growing user of 'green audits', and these also are discussed more fully later on (see p.93). Audits may be 'external' ('state of environment reports'), and involve the collation of data on the natural resource base and environmental conditions of the local authority area. The most ambitious of these have included use of public access, computerised, systems of information management; many have developed summary indicators as a means of identifying areas for targeted action and of setting benchmarks against which future change and improvements can be monitored. Audits may also be 'internal', examining a local authority's own performance in moving towards sustainable development. The same approach is also often used in the private sector and local authorities may be in a position to offer support and advice to businesses (as well as *vice versa*), based on their own experience of implementing green auditing systems.

A further aspect on which local authorities have developed expertise is that of consulting and involving local people in the design of plans and strategies for their area. In particular, 'capacity building' techniques aim to harness the energies of local people and organisations, and to aid development of the local society and economy to change in ways which are conducive to sustainability. These methods will be considered more fully later, but it is worth alluding to Arnstein's notional 'ladder of citizen participation' (see Chapter 6), which ranges from placating and informing, through involving, to empowering. Whilst direct citizen-control is perhaps unlikely, local authorities have often alleged a genuine desire for communities to develop their own sustainability strategies and to take as much responsibility as possible for implementing them. The use of round tables, networks and forums has been significantly cultivated during the Local Agenda 21 process. Many of these have drawn upon the more ambitious participation programmes associated with the plan preparation elements of town planning, though one of the benefits of the multi-disciplinary nature of sustainability planning is that it has drawn experience and skills from non-traditional areas, such as visual arts and community development. Capacity building exercises of this nature may be pivotal to re-establishing a sense of pride and responsibility in citizens.

Conclusion

Sustainability at the local level is unlikely to happen in isolation, especially in complex industrial and post-industrial economies. Whilst spontaneous enthusiasm and neighbourhood action will be essential to maintaining energy levels for the implementation of sustainability programmes, they will not by themselves form a sufficient basis. One of the main reasons why sustainable development now appears to be a durable and widely supported goal, is that it is accompanied by a large raft of commitments and actions programmes. Global agendas, continental programmes and national policies are being linked in a web which, if not exactly seamless, is at least strong enough to provide mutual support and intelligence. The connections which exist between neighbourhood action plans and wider sustainability strategies are now sufficiently visible and robust to start to give practical meaning to the maxim 'think globally, act locally'. Those responsible for devising and implementing sustainability measures at the local level need to be aware of the wider framework within which they are being facilitated and which gives them a collective purpose. In terms of information exchange, sharing experiences, building individual and organisational capacities, changing values and practices, and working towards common objectives, a continuous chain is being formed between the village green and the global common.

3

Sustainable localities

Treading lightly upon the Earth

One of the greatest challenges which humanity faces as we move into the twenty-first century is that of meeting the needs of a rapidly rising world population. However, the problem is not only one of sheer numbers of people: it is also one of limiting the scale of the impact which individuals, especially in economically advanced nations, make on the Earth's resources. One way of expressing this is that, both individually and collectively, we must learn to 'tread more lightly upon the earth'. The intensity of our impact on the earth's regenerative systems has often been severe – at times, irreversible. In our homes, workplaces, villages, towns and in our use of land and water, we must live within the regenerative capacity of natural systems if development is to be sustainable. Equally, one country or community cannot pursue apparent sustainability at the expense of another, by, for instance, subjecting it to touristic damage, by exporting its noxious wastes, or by importing goods and services which themselves plunder a host region.

This does not mean that we should retreat to subsistence levels of living. If the principles of sustainability are to be acceptable to future generations, we must acknowledge human tendencies for comfort and for capital accumulation. Indeed, many people believe that global sustainability depends on freer world trade and that only the dismantling of restrictive cartels and trade barriers can satisfy the 'equity' dimension of sustainability. Fairer trade would, in turn, need to be associated with substantial economic and structural change to benefit less developed, debtor nations.

There are presently various concepts which help graphically to portray, and perhaps even to analyse, the intensity with which we are treading on the earth. Although they have quite serious theoretical limitations as accurate methods of analysis and synthesis, they are nevertheless valuable communicative devices. The challenge is to see whether any of these alternative measures of environmental demand are both analytically sound and are usable in ways which convert people to the need to 'tread more lightly'.

Alternative measures of environmental demand

Various measures of ecological demand have been used to illustrate the ways in which human communities have affected the environment (this review draws heavily on International Institute for Environment and Development, 1995). One of the earliest attempts to express collective environmental impact was that of Georg Borgstrom (1967), who invented the phrase *ghost acres* in 1965 to describe land required to supply a country's import of food, feed and fish. Roughly speaking, developed countries appeared to require one ghost acre overseas for every actual acre in their own country. In this sense, therefore, Borgstrom estimated that they had a 'shadow ecology' of about twice their apparent one.

A currently vogue term is that of 'ecological footprint'. This describes the tendency of urban regions to appropriate the carrying capacity of 'distant elsewheres'. The notion of ecological footprints is based on the simple, but radical, observation that, whilst we are used to thinking of cities as geographically discrete places, most of the land 'occupied' (in the sense of a shadow ecology) by their residents lies far beyond their borders. This method of analysis has been pioneered in British Columbia, Canada, where it may also form a basis for future federal State of Environment reporting. An ecological footprint is an estimation of the area of land required to sustain a given community. The higher the material standard of living, the larger the footprint, and, by implication, the less sustainable the lifestyle. In principle, all uses of the earth's resources can be converted to an equivalent land area, and this can be used to estimate the land requirement of contemporary lifestyles. The impacts of individual people on the earth can be calculated in terms of their own 'personal planetoid', and this may be very large in comparison with the amount of productive land and finite resources which are readily available. An application of this method is reviewed in the next section.

The relationship between consumption patterns and the global environment has been an important component of environmental planning in the Dutch National Environmental Policy Plan. Moreover, in its report to UNCED, the Dutch Government admitted that the Netherlands economy could only sustain its current level by exploiting the 'ecoscope' (i.e. carrying capacity) of other countries. It calculated, for example, that for every hectare used in the Netherlands for dairy farming, two hectares were used in other countries. These were mainly in developing countries which exported animal feed. In those developing countries, serious environmental degradation could result from the need to set aside ever larger areas for low value cash crops such as animal feed. It also noted that, by its very nature, international trade has led countries to exploit the global ecoscope. Thus, it was concluded that the optimisation of ecological demands would require 'space' to be created for economic growth in developing countries, and this space would have to be taken from the land that industrialised countries currently appropriate for their own needs. This analysis has led to the concept of *environmental space* or *eco-capacity*.

This notion has been further generalised into the notion of a 'bubble' representing the total quantity of pollution, resource, land and sea used by a particular nation or community. The maximum size of this bubble can be compared to carrying capacity, whose use cannot be exceeded without transgressing the rules of sustainable development.

Another concept is that of the *ecological rucksack*, which refers to the total mass that each item of consumption carries with it from the cradle to the grave. In Germany, the Wuppertal Institute has calculated that, to provide the goods and services enjoyed by an average citizen, some 50 tonnes of materials have to be moved somewhere, be it in mining, earth movements for agriculture, or construction activities.

The dependence of developed nations on the resources of the Third World is a long-standing phenomenon. Some argue that each year the total ecological footprint can be interpreted as accumulating into an ecological debt owed by the developed to the developing world. In Sweden, the concept of *environmental debt* has been employed to describe the environmental burden passed from one generation to the next. It has been defined as the cost required to repair 'restorable' environmental damage, plus the cost of recurrent restoration measures.

The UK-based SAFE Alliance has taken distance from production to market of agricultural products – 'food miles' – as its indicator of the international environmental impact of British lifestyles. This emphasises the huge distances, and thus energy waste and wider impact, associated with the import of luxury (or even commonplace) items. For example, apples have to be brought 8,000 km from South Africa, shrimps 21,000 km from Bangladesh and green beans 6,500 km from Kenya. Cheap non-renewable fossil fuel energy makes intensive agriculture and long-distance transportation economically viable. Prices in shops do not, of course, reflect the full cradle-to-grave costs inflicted upon society and the environment.

As well as providing useful conceptual and communication devices, these measures help us answer some very practical questions. For instance, they can aid assessment of the impact of new technologies or policies in terms of their environmental efficiency and resource usage. They may also help evaluate the long-term effects of policies and proposals on the natural resources of a region. This may assist strategic land use planning by including analyses of the dependency of regional economies on resources from elsewhere, both in terms of their 'exported' environmental impacts and their vulnerability to sharp changes in the availability of prices of imported biophysical resources. In terms of individual products, integrative methods of analysis can help identify the effects associated with each of their lifecycle stages: development, production, distribution, use, obsolescence and disposal.

Some critics argue that these methods are too simplistic and cannot be used with any degree of accuracy. Nevertheless, their basic assumption is compelling: that the ultimate goal of sustainable development is to tread less heavily on the Earth. Whatever their computational limitations, they convey very graphically our current and possible future sustainability.

Sustainable regions

Although above the scale of the 'local', the region serves as an important contextual unit of environmental analysis. The factors which historically have defined the boundaries of regions may now make little sense in terms of their administration and management. In environmental terms, it may make much more sense to consider regions on the basis of of their river catchments or biological habitat types. For the purposes of sustainable development, this could provide a framework for the integration of environmental work programmes between public agencies and private land managers. Considerations such as these have led some environmentalists to argue the case for 'bioregions' as a unit of administration and government, namely, areas possessing internal unity of land and water resources which can be governed on a sustainable use basis. It is, of course, very unlikely that statutory administrative areas will be reorganised on this basis alone. However, notions of sustainability are likely to condition the way in which we think about regions, and the concept of 'bioregions' is useful in shaping our ideas about the resource consumption patterns of large administrative areas.

Nijkamp et al (1992) identify three approaches to defining the sustainability of development at the regional scale. One is a fairly conventional environmental planning model of minimising conflicts between resource-using activities, to enhance socio-economic conditions in the present and yet bequeath an environmental estate to the future. A second perspective draws more explicitly on the inter-regional input-output analysis approach, which views a region as an open spatial system. Within this, sustainability might be construed as involving a long-term balance of flows of regional imports and exports, both ecological-physical flows and trade-monetary flows. The physical flows are determined not only by cross-boundary flows of ground water, surface water and air, but also by socio-economic activities such as trade, capital flows and migration. Although this is a sophisticated and powerful basis for analysis, Nijkamp et al consider its emphasis on 'balanced flows' to be too limited. Their third – and preferred – perspective is to view regional sustainable development as development which ensures that the regional population can attain an acceptable level of welfare, both at present and in the future, and that this regional development is compatible with long-term ecological viability.

Thus they see regional sustainable development as needing to fulfil two goals:

- it should ensure an acceptable level of welfare for the regional population; and

- it should not be in conflict with sustainable development at a supra-regional level.

These do not necessarily imply that all regions need to be internally self-sustaining. Indeed, it is possible that some regions (and, by implication, smaller areas of town and country) may need to make 'sacrifices' of welfare and environmental quality, in order to assist sustainability at supraregional or global levels. Such inequalities may need to be addressed by specific compensatory mechanisms.

The most striking attempt to determine the ecological footprint of a region has been that of Rees and his colleagues (e.g. Rees, 1992) for the Lower Fraser Valley in British Columbia, Canada (note that this approach can be used at all levels from the national to the personal and is also currently generating interest at the municipal and neighbourhood levels). The Valley is an area of over 4,000km², comprising a population of some 1.7m people and it enjoys a high standard of living. The likelihood is, therefore, that its ecological impact is disproportionate to its spatial or demographic scale. The method used to calculate its sustainability was an ecological accounting approach using land-area as the main biophysical measurement unit. The analysts presumed that the region's activities could be reflected in its consumption of resources or production of wastes. These could be quantified as:

- the land required to produce farm or forest goods, or their derivatives, and
- the land or water space needed to assimilate (without harm) the wastes produced by the region.

A specific problem arises in relation to how to include the use of fossil fuels and their derivatives (such as plastics) which cannot readily be measured in spatial terms. This was circumvented by assuming that ethanol (produced as a biofuel from crops) could potentially be used as a substitute for fossil fuels. The equivalent land area of fuel crops could then be taken as a proxy for fossil fuel consumption.

Each aspect of production or consumption is thereby defined as a category of social or economic activity. Adding up the land requirement of all these categories yields an aggregate or total area which represents that region's 'ecological footprint' on the earth (Figure 3.1). This may also be thought of as an 'appropriated carrying capacity' – in other words, the amount of resources which are required to support a given population both from within its host region and from beyond its boundaries. Clearly, in the case of non-subsistence economies, this population is likely to rely on other regions to supply some of its needs, so that part of the carrying capacity requirement must be 'appropriated' from elsewhere.

The ecological footprint has been defined as the aggregate land (and water) area required by people in a region to provide continuously all the resources they presently consume and all the wastes they presently discharge. For purposes of calculation, this area is subdivided into various categories.

Table 3.1   Generalised matrix for calculating ecological footprint

Units = ha/capita	Energy	Degraded land	Garden	Crops	Pasture	Forest	**Total**
food (vegetarian, animal products)	X_iY_i ...	....	....	....	....	X_iY_n	Σ
housing (construction, operation)							Σ
transport (private cars, etc.)							Σ
consumer goods (e.g. clothing)							Σ
resources in services received (e.g. education)	$X_nY_{i...}$	....	....	....	....	X_nY_n	Σ
Total	Σ	Σ	Σ	Σ	Σ	Σ	ΣΣ

Source: based on Wackernagel and Rees, 1996

Figure 3.1   Depiction of an ecological footprint

Source: Jannson et al, 1994

Predictably, determination of the ACC often has to make some heroic assumptions. Having estimated the inputs and outputs, however, a matrix can be produced, with the consumption categories (e.g. food, housing) arranged in rows, and the land use categories (e.g. energy, pasture) in columns (Table 3.1).

Using this approach, it has been calculated that the footprint of the experimental area in the Lower Fraser Valley (British Columbia) is some twenty times the size of the region's area. Clearly, modern societies need to import and export goods, and distant exporters or importers may benefit greatly from the region's consumption patterns. It would be pointless, therefore, to argue for a direct equivalence between a region's area and its ecological footprint. Nevertheless, this analysis does raise thought-provoking questions about the sustainability of particular regional economies and societies. If everyone in the world possessed similar personal planetoids to the people living in this region, it would (according to these particular calculations) require three planet Earths to sustain them.

Sustainable settlements

Recognition of the relationship between buildings, settlement patterns and resource consuption is hardly new. Yet it is only comparatively recently, with the emergence of sustainable development as a politically cogent issue, that it has started to colonise mainstream planning doctrine. Thus, an increasing amount of official attention has been given to the ways in which alternative patterns and specifications of roads, houses and workplaces can help limit pollution and energy use. At the UK level, the Sustainable Cities Network has been supported by the Economic and Social Research Council to co-ordinate and disseminate research on urban sustainability. Internationally, the International Council for Local Environmental Initiatives has promoted 'capacity building' within local government structures as a means of pursuing sustainability at municipal level, and has partially sought to address this through a research network of Model Communities (see Chapter 5).

As Owens (1992) notes, sustainable urban development is probably a contradiction in terms as, by definition, urban areas require the resources of a wider environment for their survival. During most of the twentieth century, the relative absence of energy constraints has permitted increasing separation of activities and outward spread of urban areas at decreasing densities. Urban structure is itself an important determinant of energy demand, especially for transport, space-heating and ventilation. Since, in the UK, space-heating and transport account for well over half of delivered energy needs, the importance of municipalities and their administrations in pursuing more sustainable pathways is self-evident. It has been estimated that as much as seventy per cent of delivered energy may be susceptible in some way, and at some time to being influenced by land use planning mechanisms. Given that building and transport systems, once constructed, are relatively permanent and have limited capability to respond to short-term fluctuations in economic pressures for short-term energy conservation, it would seem sensible to incorporate energy efficiency into urban design as a safeguard against future deterioration in energy supplies or market prices.

Owens' analysis identifies various key areas of interaction between energy and planning. One of these is at the micro-level – of built form and design for passive solar gain – where style and density of houses, coupled with site environmental factors, can exert considerable influence over energy use. For example, detached houses can theoretically (though rarely in practice) require three times the energy input of equivalent accommodation provided in an intermediate flat, and this should be a significant consideration in the light of likely future needs of smaller dwelling units (e.g. for retired people, starter homes, sub-nuclear families). Optimum use of solar gain and microclimatic conditions can further substantially reduce the need for space heating (or cooling) of buildings from conventional sources. Combined heating and power (CHP) systems are also viable in carefully conceived schemes and can result in major efficiency gains: conventional power stations convert primary fuel into electricity with a maximum efficiency of about thirty per cent, whereas CHP (using heat produced during the generation process for space and water heating) can raise this to around eighty per cent. CHP is one of a number of ways in which energy distribution can be planned into urban design, and some pilot schemes are underway to explore its application in practice. Carefully designed urban form can also substantially reduce the need for movement. Whilst, for example, modifying the density of settlement and the degree of mixing of different land uses will not automatically result in reduced energy consumption, it may well minimise physical separation between people and their destinations. This, in turn, should create opportunities to reduce travel needs and increase the feasibility of making trips by cycling and walking. Equally, the location and form of urban design can affect choice of transport mode, and thus of the types of travel likely to be used.

Appropriate planning policies could therefore include:

- discouragement of dispersed low-density residential areas or any significant development highly dependent on car use;
- some degree of concentration, though not necessarily centralisation, of activities;
- integration of development with public transport facilities and the maintenance of moderately high densities along transport routes; and
- if transport networks do not already exist, planning them in an integrated way with the development of land.

The establishment of all of these as policiy options is leading to the creation of an increasingly 'standard' wisdom. However, generalisations can be misleading: each urban area will have particular characteristics which promote or preclude certain energy-efficient options. It could well be argued that in earlier decades, when there was an emphasis on the mass production of goods and services, and an associated concentration of activity, there was greater scope for journey minimisation. This type of activity pattern is often referred to as 'Fordist'. Advanced economies are increasingly organized on a 'post-Fordist' basis, in which flexible industrial practices lead to a greater dispersal of activity, often indirectly accompanied at a personal level by complex individual time budgets. In this situation, working and domestic patterns can militate against the simple journeys for which public transport can most readily cater.

Physical considerations, though, do not of themselves consititute a basis for sustainability and it is clear that social capacity building is of equal importance. A top priority in the quest for sustainable urban development is the creation of viable political and institutional systems capable of framing broadly based strategies, programmes and policies. Haughton and Hunter (1994) suggest that a sustainable city is one which its people and businesses continuously endeavour to improve their natural, built and cultural environments at neighbourhood and regional levels, whilst at the same time working in ways which always support the goal of global sustainable development.

Agenda 21 has identified the major physical environmental issues facing urban settlements world-wide. These include improved planning and management, integrated provision of environmental infrastructure (for basic water supply and sanitation), the promotion of sustainable intra-urban energy and transport systems, anticipatory planning in disaster-prone areas and urban health. It is acknowledged that effective attempts to address these issues must be accompanied by participatory approaches and by enhanced capacities within local government. In more affluent countries, urban sustainability is likely to place greater emphasis on quality of life: the Green City programme for New York, for example, advocates more grass roots involvement, cross-generation environmental education and open space provision, as well as 'environmentally sound development' embracing air, water, energy and waste management.

Planners in Britain have been encouraged to accommodate more sustainable development patterns in their forward plans, which is a remarkable shift of emphasis from the philosophy of land use planning which gave primacy to the car. An official aim of planning is thus now to guide new development to locations which reduce the need for car journeys and the distances driven, or which permit the choice of more energy-efficient public transport (DoE, 1993a, 1994a). In addition, cycling and walking may be encouraged by providing safer and more attractive routes – an option which has been given a considerable boost by Millennium funds to speed up the provision of a national cycle network. By the same token, provision of transport infrastructure can itself influence future settlement patterns.

In broader terms, planners are encouraged to make full and effective use of vacant land within urban areas. At its worst, this approach can lead to 'town cramming', in which valued local spaces and views are lost to incremental 'infill' and 'backfill' developments. However, it can also make use of reclaimed 'brownfield' sites and under-utilised land which can prevent the loss of high-

Table 3.2 Sustainability considerations in transport planning

Housing
- allocate the maximum amount of houses to existing larger urban areas where facilities and a range of transport provision can be easily accessed;
- promote land for housing in locations capable of being served by public transport if it cannot be found in central locations;
- avoid significant incremental expansion of settlements if likely to lead mainly to travel by car;
- avoid sporadic housing development in open countryside;
- avoid development of small new settlements, especially if unlikely to be self-contained or served by public transport;
- re-use land, recycle buildings and, where possible, promote mixed-use development;
- set standards to maintain and, where appropriate, increase existing densities; and
- concentrate development near public transport centres/corridors.

Employment
- move towards a better balance between employment and population in order to enable people to live near their work;
- focus uses which generate intensive travel (e.g. offices) in areas well served by public transport; and
- avoid major developments in locations not well served by public transport or otherwise not readily accessible to a significant local workforce.

Freight
- designate sites for distribution and warehousing, especially of bulk goods, in sites accessible to trunk roads, wharves, harbours or railway sidings; and
- maximise proportion of materials moved by rail or water.

Retail
- maintain and revitalise existing central and suburban shopping centres;
- encourage local convenience-shopping, in areas which are attractive and readily accessible on foot or bicycle;
- avoid sporadic siting of comparison-goods shopping units out of centres or along corridors; and
- where feasible, provide for both local shopping and residential uses in large new developments.

Leisure, tourism and recreation
- concentrate features in locations well-served by public transport, including cinemas and theatres in central locations, to maintain town's vitality;
- maintain and encourage the provision of local leisure and entertainment facilities. and
- make provision for attractive and accessible local policy areas, public open space and other recreational facilities.

Education and other public facilities
- ensure facilities with wide catchment areas are accessibly located with respect to public transport; and
- prefer urban settings for university expansion.

Promote complementary transport measures
- increase the relative advantage of means of travel other than the car, especially walking, cycling and public transport;

- reduce dependence on the private car, and discourage its use via restrictions on parking facilities; and
- increase the competitiveness and attractiveness of urban centres against peripheral development.

Source: derived from PPG 13, DoE, 1994a

---

amenity greenfield sites in peripheral locations and which will render users heavily dependent on cars for mobility. In order to reduce the need always to undertake journeys by private transport, there is considerable scope for locating new development close to public transport networks, ensuring for instance that it is situated near railway stations with spare capacity, and for better locating public transport facilities generally.

The location of new retail and service facilities is also being reappraised, and the predominantly 'market led' approach to out-of-town complexes is starting to be challenged. These types of land use *attract* trips and, if located on the edge of cities or near motorway junctions are bound to encourage car-borne users. Conversely, if they are channelled to town centres, they have the potential to revitalise urban economic fortunes and to make better use of public transport. Equally, new housing can be related to local facilities, so that cars are not always necessary to access schools, shops, and so forth. Of course, there are disadvantages to the promotion of town centres, particularly in terms of economy, convenience and congestion, and the balance is a delicate one to strike. It may be necessary to limit town centre parking facilities, either by price or capacity, to deter car use, though this must be weighed against the possible adverse reaction of consumers. It has been observed that, whilst planners cannot compel people to lead more energy-efficient lifestyles, they can encourage development patterns which at least make such choices possible (Table 3.2).

Energy consumption within buildings may also be reduced by a variety of increasingly well-tested mechanisms, though it is naturally easier to achieve this with new, purpose-built buildings than with ageing, inherited ones.

Without wishing to detract from the UK's widespread achievements in domestic and industrial energy conservation, it is useful to consider the approach in Denmark, where there has been signal success. In broad terms, despite an increase in housing area of about 30% since the beginning of the 1970s, space-heating now only accounts for about 28% of total energy consumption compared with 39% in 1972 (Danish Energy Agency, 1993). Relatively straightforward measures, comprising extensive energy-saving devices in old buildings and stricter regulations in new ones coupled with fairly high energy prices, have permitted energy consumption per m^2 of heated area to be reduced by 45% during this period. Measures involve thick insulation (as much as 300mm against the roof), energy-efficient windows, utilisation of passive solar heating, low-temperature district heating and the operation of advanced standards of ventilation and other installations. Standards are being continuously upgraded, and regulations introduced in the early 1990s have set upper limits for

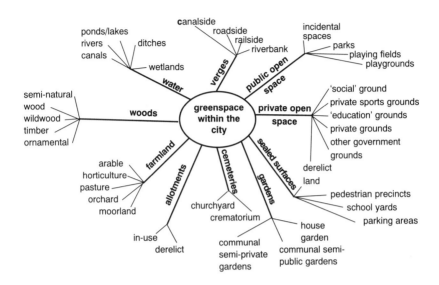

Figure 3.2 Elements of urban greenspace

Source: Beer, 1994

total energy demand for heating and ventilation. The principle of imposing an overall limit, rather than prescribing specific measures, enables architects to adopt more flexible approaches. Low-energy housing can thus incorporate innovative design and high levels of comfort. Many large-scale (i.e. industrial and commercial) consumers have also introduced energy management techniques to keep track of their energy consumption. Coupled with this are co-ordinated campaigns to introduce energy management into all large state-owned buildings and to promote the purchase of energy-efficient equipment. More extensive experiments are underway in two selected towns (Toftlund and Bredstedt) to establish whether the environmental impact of the energy sector can be halved. Early projects in these towns included a biogas plant and flexibly controlled outdoor lighting schemes.

A further element of 'sustainable' settlements is their greenspace. Open spaces comprise a complex structure of formal and informal sites, some of which were specifically designed for public recreation, and others which accidentally have acquired an amenity or environmental function (Figure 3.2). A major study in Sheffield has greatly enhanced our understanding of the ways in which the links between a city's built-up areas and open spaces influence its capacity to become more sustainable. This study confirmed that, potentially, green spaces can not only support recreation and nature, but can also lead to

Table 3.3   North Sheffield sustainable environments and the role of greenspace Project (University of Sheffield)

Abiotic
microclimate variations
topographic variations
local surface drainage and water quality

Biotic
% of surface cover by woodland trees, individual trees and large shrubs
% of surface cover by mown grass, rough grass, scrub
% of surface cover by water, wetlands
variety of plant species
relationship of site to Sheffield's green corridor plan
sites of special ecological interest – location, size and shape

Social
population distribution by age, sex, family type
poverty indicators
% unemployed
% of single parent families
housing tenure
housing for sale
level of vandalism
subjective assessment of feeling of 'safety'

Land Use
use of building
age of building
condition of building
block height, length and width
size of front and rear gardens
number for sale
transport
width of streets and verges
width of other tracks and verges
public transport accessibility
woodlands by type, size and location
agricultural land
derelict land
% sealed surface (paved and built over areas of land)

Source: Beer, 1994

improved water and air standards and general enhancements to quality of life. A better understanding of their roles and dynamics may also help target financial resources for regeneration and community action.

Variations in the amount, type and structure of greenspace appear to make some urban layouts more efficient than others in the way they interact with the environment. Having entered extensive data on the abiotic, biotic and social features of Sheffield (Table 3.3) into a geographic information system, the project was able to examine the the city in terms of its 'urban landscape structure zones'. These revealed how one area differed from another and why some zones more than others appeared to achieve the levels of sustainability implicit in the city's planning policies. The expectation is that performance criteria can be established for a range of environmental attributes and these, coupled with the widespread involvement of local communities, may lead to novel approaches to environmental management. These approaches might include:

- local water-management using the house, garden and local open space to manage water;
- ameliorating heat loss in buildings by increasing tree cover in exposed locations;
- increasing tree cover to ameliorate outdoor climate;
- increasing the range of habitats throughout the area to enhance the level of biodiversity;
- growing biomass in extensive open space for use in small community combined heat and power schemes; and
- developing effective local composting of biodegradable garden and household waste.

Innovations of this kind have already been incorporated into urban design in some settlements. For instance, in Örebro, the suburb of Ladugårdsängen has been planned on the basis of ecological principles, such as extensive cycleways, mixed-use neighbourhoods, rainwater collection, waste separation and decreased waste collection frequencies (Gustafsson, 1994).

One of the most persistent debates surrounding sustainable urban development has been the desirability or otherwise of promoting centralised and high-density settlement patterns. Most literature suggests clear advantages for centralised cities, including some types of linear city (i.e. settlements configured along an axis of high quality, rapid mass transit services). Indeed, it is apparent that certain compact or linear patterns of development are intrinsically more energy efficient than other options, in terms of trip generation and scope for public transport and shared heating networks. Breheny and Rookwood (1993) also note that the conventional wisdom of high-density, mixed-use (i.e. mixing residential and commercial/light industry uses to locate workers close to workplaces) cities has spread to the EU. *Europe 2000* (Commission of the European Communities, 1990a), for example, views this type of settlement as being energy efficient and leading to higher quality of life, because the conurbation has less 'urban sprawl'.

However, there is a strong case that this is contrary to human behaviour and preferences. Historical attempts to work against human nature have been doomed to failure: town planning, in particular, has often been disastrous when it has attempted to reverse fundamental attitudes and processes. This is a significant consideration at a time when we are adjusting our inherited settlement patterns from the relatively brief historical episode of the industrial revolution and refocusing our population into the more organic pattern of county and market towns. It is likely that assumptions about re-urbanisation or reversion to simple inter-city transport patterns, susceptible to mass transit

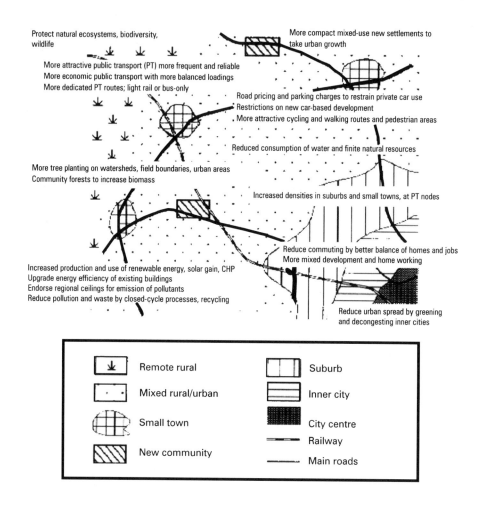

Protect natural ecosystems, biodiversity, wildlife

More attractive public transport (PT) more frequent and reliable
More economic public transport with more balanced loadings
More dedicated PT routes; light rail or bus-only

More tree planting on watersheds, field boundaries, urban areas
Community forests to increase biomass

Increased production and use of renewable energy, solar gain, CHP
Upgrade energy efficiency of existing buildings
Endorse regional ceilings for emission of pollutants
Reduce pollution and waste by closed-cycle processes, recycling

More compact mixed-use new settlements to take urban growth

Road pricing and parking charges to restrain private car use
Restrictions on new car-based development
More attractive cycling and walking routes and pedestrian areas

Reduced consumption of water and finite natural resources

Increased densities in suburbs and small towns, at PT nodes

Reduce commuting by better balance of homes and jobs
More mixed development and home working

Reduce urban spread by greening and decongesting inner cities

⊻ Remote rural	Suburb
Mixed rural/urban	Inner city
Small town	City centre
New community	Railway
	Main roads

Figure 3.3   The 'social city region'
Source: Breheny and Rookwood, 1993

provision, offer at best only a partial truth. Sustainable pathways may well emerge by bringing work closer to people (for example, by teleworking in the very broadest sense) and by acknowledging the cultural facilities which only cities can provide.

It is interesting, therefore, to note the analysis of the Town and Country Planning Association (Blowers, 1993), which argues that compact cities are quite contrary to deeply embedded geographical trends of 'flight from the city'. The TCPA therefore suggest a 'social city region' (Figure 3.3) in which all types of settlement pattern from city centre to remote rural are complementary and mutually reinforcing. They propose that we should make a series of changes in the ways we use environmental resouces if we are to achieve future sustainability (Table 3.4); this can be linked to a checklist of criteria against which progress towards sustainability can be monitored (Table 3.5). One of the most revealing features of this analysis is that different components of the social city region (city centres, inner city, suburbs, small towns, new communities, mixed urban-rural areas and remote rural areas) can be scored against the checklist of criteria for their performance in contributing to environmental sustainability. Making some perfectly reasonable assumptions about travel, recycling, 'alternative' energy production, economic diversification, etc., each of these areas can substantially improve its contribution to sustainable development.

Table 3.4   The 'social city region': changes needed for future sustainability

Natural resources
- increased biological diversity, including positive measures for encouraging wildlife;
- big increase in biomass (trees and other green plants) in both town and country;
- replacement instead of depletion of groundwater reserves and good quality topsoil; and
- much greater use and production of renewable materials in place of scarce finite ones.

Land use and transport
- shorter journeys to work and for daily needs;
- much higher proportion of trips by public transport;
- more balanced public transport loadings to minimise fuel consumption;
- greater local self-sufficiency in non-speciality foods, goods and services; and
- more concentrated development served principally by public transport.

Energy
- greatly reduced consumption of fossil fuels;
- increased production from renewable sources, e.g. sun, wind, tides and waves;
- reduced wastage by better insulation, more use of CHP, local power generation; and
- form and layout of buildings, better designed for energy efficiency.

Pollution and waste
- reduced emission of pollutants, especially from industry, power stations and transport;
- comprehensive measures to improve the quality of air, water and soil;
- reduction in total volume of waste stream;
- greater use of 'closed cycle' processes; and
- much greater recovery of waste materials through recycling.

Source: Breheny and Rookwood, 1993

Table 3.5   Checklist for monitoring progress towards future sustainability

---

Checklist for monitoring progress

1. Pollution reduced by:
- establishing the environmental capacity of the region for emission of pollutants;
- refusing permission for any development that would result in the total volume of emissions exceeding the regional capacity; and
- setting up inducements and penalties to cut existing emissions.

2. Natural resources conserved by;
- encouraging rehabilitation rather than redevelopment;
- stimulating regional production of renewables to replace finite non-renewables; and
- adopting conservation measures to save topsoil.

3. Total volume of waste stream reduced by measures such as:
- reducing business rates for firms using 'closed cycle' processes;
- introducing graduated charges for waste collection.

4. Increased recycling of most waste materials including:
- recovery of scarce inorganic materials for re-use; and
- composting of organic wastes.

5. Reduced energy consumption and increased percentage from renewables by:
- programme for raising energy efficiency of all buildings to at least minimum sustainability standards;
- increased use of solar gain;
- greater use of combined heat and power systems; and
- development of wind farms and wave power.

6. Major increases in biomass, both urban and rural, by:
- more community forests and other rural tree-planting;
- protection of existing urban open space and creation of new open space in areas of deficiency;
- additional urban tree-planting and other green vegetation;
- gardens on flat rooftops; and
- more green areas in new development projects.

7. Regional water supplies augmented and consumption reduced by:
- tree planting to maximise rainwater retention in watersheds;
- metering consumers with graduated charges favouring low consumption;
- applying 'closed cycle' methods to water use;
- separating 'grey' water for filtering and return to groundwater reserves; and
- reducing urban run-off by use of more permeable paving, providing natural channels and lagoons in place of closed drains.

8. Urban decentralisation and dispersal reduced by:
- greening and decongesting inner cities;
- making inner-city housing more attractive by eliminating excessive densities, designing for 'defensible space';
- increasing average densities in city suburbs and small towns and;
- using more concentrated forms for new development.

Source: Breheny and Rookwood, 1993.

At the level of the individual settlement, innovative work has taken place in Canada on the Green Communities and Healthy Communities Initiatives

(Gordon, 1994). The GCI was initially the idea of the Ontario Round Table on the Environment, and it commenced at the end of 1992, aiming to:

- reduce pressures on the environment;
- help home owners save money through reduced utility bills;
- help local government and business save money through reducing demand for new infrastructure investment (e.g. new water supply and sewage treatment facilities, new electricity generation and distribution systems);
- stimulate the demand for locally-produced environmental goods and services; and
- create new jobs for local tradespeople and small businesses.

An early example was Guelph 2000, an initiative whose initial attraction to the municipal government was that it helped them defer major capital expenditure on new infrastructure. One of its key components was a 'Home Green Up' programme, which entailed home visits by advisors to assess energy and water use patterns and opportunities for households to reduce wastes. Both initiatives are now well-developed across Canada.

At a more conceptual level, the ecological approach to settlement design has been related to three basic elements, namely, flows, areas and participants (ter Heide and Berends, 1994). Settlement 'flows' comprise water distribution, traffic circulation, waste recycling and building processes. 'Areas' are the scales of planning at which ecological strategies can take place. For example, conurbation-wide planning may focus on transportation structures and open space networks; neighbourhood scale planning may be more concerned with landscapes and facilities; whilst existing built-up areas may require regeneration of their ecological infrastructure. The citizens, or target groups, for whom planning takes place are viewed as 'participants' in the process. A key issue is to ensure that strategies do not consider solely the interests of the more obvious target groups, but are also sensitive to cultural diversity and intergenerational equity. This is likely to require new technical skills on the part of planners.

The most explicit application of the flows-areas-participants framework has been in the Dutch 'Ecopolis' programme (Tjallingii, 1995). The approach is conceptually somewhat impenetrable and, for simplicity, is here distilled into two essential elements. First, each component is related to a main objective, so that the city becomes more responsible in its management of flow-chains, its areas become more culturally and ecologically vital, and its citizens are encouraged to be more participative in environmental strategies (Table 3.6). Thus:

- the regulation of flows entails a responsible approach to resource use, aimed particularly at the source of the flow chain, leading to an emphasis on 'chain management';
- the planning of built-up and open areas aims to create a 'living' city and is dependent on a coherent spatial planning policy; and

- a participative approach is required to influence the lifestyles and economic activities of different target groups.

Second, management of change in this complex type of situation cannot rely on a blueprint approach, but involves a package of policy instruments directed at goals and objectives. These include measures which are:

- influencing (information and persuasion);
- facilitating (physical facilities);
- stimulating (subsidies and financial incentives);
- repelling (levies and liability);
- limiting (restrictions on physical capacity, quotas); and
- integrating (laws and regulations).

Table 3.6   Conceptual framework for the 'Ecopolis' programme

Focus	'Responsible' city *Flows*	'Living' city *Areas*	'Participating' city *Participants*
Socio-economic objective	production of goods and services; quality of life	usefulness; attractiveness	prosperity; well-being
Problems to be addressed (examples)	resource depletion; pollution; disturbance	health; loss of local distinctiveness in biotic environment	alienation; indifference
Objective for 'Ecopolis' planners	sustainable flow-management; contol of renewable resource quality	sustainable use of areas	community commitment to ecological management
Policy goal	chain management	area management and planning	target-group policy

Source: based on Tjallingii, 1995

A great deal of debate has surrounded the ways in which sustainability plans and programmes should be implemented at the city level. Traditionally, city planning has been based on top-down rationalistic approaches which tend to exclude substantial citizen involvement. At best, these have involved consultative (usually tokenist) methods of obtaining public viewpoints. Although rational approaches are (in rhetoric, at least) becoming less fashionable, they do assist the operation of certain indispensible central technical facilities. The opposite trend has been to encourage bottom-up, grass roots action, which may lack focus and even invite the emergence of certain repugnant 'unreconstructed' attitudes. Advocates of the Ecopolis approach thus favour a 'sandwich strategy' in which the top layer of a city or sub-regional authority creates favourable economic and organisational conditions, and the bottom layer of citizens and businesses develop practical measures. An intermediate layer of government is then responsible for implementing neighbourhood and settlement-level initiatives.

Sustainable countryside

Rural areas are distinguished from urban areas by their much lower density of population and relative predominance of natural environmental processes. Rural environments tend to be more heavily dependent on natural bio-geochemical cycles and visual amenity. They are thus characterised by biological productivity (in the form of farming, forestry, wildlife and fisheries), recharge of water supplies, availability of non-renewable mineral and energy stocks, and 'positional' amenity goods. These last comprise assests which obtain their value from a distinctive location, such as one which is especially scenic or suited to recreational pursuits.

Rural areas are also characterised by small and medium-sized settlements, and these are increasingly associated with a range of manufacturing and service industries. They also possess some of our most valued cultural assets, including quintessential buildings and neighbourhoods, and are thus integral to the maintenance of a sense of 'local distinctiveness'. Rural sustainability is thus multi-faceted and incorporates:

- sustainable employment and economic performance;
- sustainable communities;
- sustainable recreation and tourism; and
- sustainable natural capital.

The 1995 White Paper on Rural England (DoE/MAFF, 1995), not surprisingly, observed that national policies for the countryside must be founded on the principles of sustainable development.

Much of the debate about sustaining the countryside has centred on retaining and invigorating rural populations. The Rural Development Commision (1995) has argued that economic and social development is vital if the countryside generally is to adapt to the major changes which are occurring and to avoid ossification. Many of the remoter rural areas, upon whose continued existence the sustenance of rural positional goods depends, have an urgent need to diversify their economies away from the traditional primary industries. One of the elements of sustainability planning must thus be to promote sustainable development in rural areas by encouraging the co-location of jobs, housing and services in villages and small towns. This is reckoned to be resource efficient, insofar as it reduces the need to travel, and results in the re-use of buildings and reclaimed land. It can also help support communities in places where it will be necessary to secure the use of renewable sources of energy and raw materials.

The RDC further echoed a widely felt concern at official forecasts of the numbers of new households to be constructed, and at the consequent pressure being placed on the countryside to accommodate new building. In the light of this, it considered that strategic decisions were needed on the balance of conservation and development in rural areas. The RDC likewise enjoined debate on whether reliance for built development in the countryside should be achieved purely by expanding some small towns and villages, or whether completely new communities should be created. One particular opportunity which now exists is to re-use some major redundant public sector sites, such as defence bases. The quality of building in villages, in terms of scale, layout and design may also help sustain their qualities which contribute to local distinctiveness.

Economic diversification in the countryside has been officially encouraged (DoE, 1992b), but problems remain. Some planning authorities try to confine new workspace development to the larger country towns, whereas *in situ* expansion of businesses in villages may be more sustainable (though certainly often more controversial). Equally, the provision of housing which is affordable to young locals could help recapture the sense that villages are places where people live and work and, in a sense, therefore, make them more sustainable than if they languished as 'dormitory' retreats.

Owen (1996) affirms forcefully that sustainable development in the countryside is about much more than the protection of ecological and cultural resources. Thus, he argues that sustainability plans should be multi-faceted, incorporating factors such as:

- employment-led rural regeneration responding to local needs;
- encouraging the development of well-designed affordable housing where local need can be demonstrated;
- protecting and responding to the character of the natural environment, whilst seeking to derive environmental benefits for the local community.

Further, there is a case for policies which encourage self-reliance and urban containment, whilst discouraging the modern shibboleth of trip-minimization. He refers to the rejection, by a planning authority, of an eminently 'sustainable', self-sufficient house, which conflicted with a policy to locate new development only in well-serviced settlements.

The issue of sustainability in agriculture is enormously broad and, for manageability, the brief account here is restricted to a few salient features of 'first-world' land use (for a more thorough review, see Winter, 1996). Sustainability in farming is a function both of the farm and the farmer. The farm will be characterised by its physical environment, its enterprises and its fixed and mobile assets, and all these will display limits which could make the farm inherently sustainable, or sustainable only with certain types of policy support. Studies of land use change in the countryside have often ignored the farmer, but it is the farmer's attitudes and actions which determine environmental changes, and the ability of farming families to cling on which sustains the viability of many local rural economies and communities.

Modern agriculture has been accused by environmentalists of being unsustainable for many reasons, including pollution by biocides, diffuse pollution of watercourses and aquifers by nitrates, and heavy reliance on chemical and energy subsidies. The effect of these clearly varies according to local environmental conditions: nitrate pollution is, for example, only a serious problem in areas with particular hydrogeological characteristics. Equally, farmers have become heavily indebted and been buffeted by waves of policy changes which have made it difficult for them to manoeuvre. Many personality factors and business circumstances influence the ways in which farmers will respond to new policy measures. Recent studies have tried to model the complexity of physical, economic and human considerations and the way they interact. The NELUP study (O'Callaghan, 1995), for instance, integrated a large quantity of data on biophysical characteristics, with human and economic data, to produce a decision support system indicating the likely nature and environmental significance of farmers' responses to policy changes.

Agricultural policy has, in recent years, started to shift away from its 'productivist' basis, reliant on protectionist and subsidised policy support, to one of limited assistance and a gradual drift towards open market conditions (i.e. world food prices). This is a difficult and controversial task, both because of the dependence on policy support which farm businesses had developed over the years, and the political strength of the farming lobby. Some of the policy shifts are purely economic (levels of price support on grain and livestock), whilst others have a strong environmental bias ('environmentally sensitive areas', habitat payments, nitrate sensitive areas, organic aid, public access). These will, it is hoped, also depress production of surplus foodstuffs, or at least discourage further land use intensification. In some policies, there is a small degree of articulation between environmental and output objectives ('cross-compliance'), such as in the moorland scheme or, more abstrusely, in set-aside. However, it is enormously difficult to steer a course which achieves sustainability in policy support terms (i.e. levels of financial support which the country can sustain long-term), farm business terms (i.e. financial stability), environmental terms (i.e. minimal damage to wildlife, soil, scenic and water resources) and social terms (i.e. farmers and farm workers contributing to local social networks).

Forestry is the second major user of rural land, and is similarly multi-faceted in terms of its relationship to sustainability. We have previously noted that forestry, in particular, has been associated with notions of sustained yield and sustainable management for a long time (see p.10). However, important additional aspects are now starting to emerge. Latterly, there has been greater recognition of the contributions which forestry might make to controlling atmostpheric CO_2 levels (woodland is a carbon sink), diversification of farm economies, and short-rotation coppice for energy production. Woodlands can also contribute to local sustainability, where there is genuine neighbourhood involvement in the design, implementation and management of community woodlands. Whilst in many parts of the 'third world' there is a pressing need for the establishment of local woodlots, for instance as fuel or constructional sources, it is often a far more difficult task to convince people in the 'first world' of the need for reforestation on either large or local scales. Once more, there is a difficult balancing task in creating forests and woodlands which contribute to sustainable development in multiple ways.

The landscape quality of the countryside may also be related to notions of sustainability. Our understanding of the value of landscapes tends to change markedly over time. Not only are our aesthetic tastes and perceptions of beauty quite volatile, but also our appreciation of the functions performed by landscapes changes, and landscapes are themselves inherently dynamic in both cultural and natural environmental terms. It therefore difficult to identify a single diagnostic feature of 'sustainability' with respect to landscape, but it is essential to pursue this issue as our perception of landscapes is profoundly influential on our general custodianship of the countryside.

One way in which this aspect of sustainability might be approached is through the discipline of landscape ecology (e.g. Selman, 1993, 1996). This refers to the ecology of large tracts of land as analysed in terms of their wider patterns and processes. Although much of the subject matter is controversial, it may be assumed that severe landscape fragmentation is likely to lead to a reduction in biodiversity because it limits the opportunities for species to migrate between viable habitats and for local populations of the same species to inter-breed. Thus, a landscape system which contains certain habitat structures may possess self-organising properties, and in that sense be sustainable. It may also be associated with types of visual complexity which are appealing to humans, though this topic requires much elaboration. Also, some of our most biodiverse landscapes are those associated with traditional (and perhaps obsolescent) farming systems, and their sustainability will depend on the retention of cultural systems of land management. In such cases, agricultural support or conservation policies may be geared to the production of amenity rather than the production of maximum food.

Finally, the landscapes which prove most sustainable will often be those which people want to see, and thus sustainable development may require the active participation of local people and visitors in designing landscape futures. Various techniques have been used to help observers to visualise the outcomes of certain planning and policy options on the landscape by combining economic and ecological modelling techniques with digital or manual methods of generating visual displays. These are more fully reviewed in Chapter 4.

A further activity which has greatly affected the sustainability of local environments, economies and communities is tourism. From originally being seen essentially as a business proposition, tourism is now viewed more sensitively as a balance between the place (environment), visitor and host community. Consequently, recent effort has been invested in the pursuit of 'greener' tourism (though an extreme view would be that a green tourist is someone who stays at home). Green tourism assumes that if benefits are to be maintained in the long term, activities need to be inherently sustainable rather than 'boom-and-bust'; thus short-term exploitation of an area which leads to the attrition of natural or cultural assets is avoided.

Table 3.7 Principles for sustainable tourism

The environment has an intrinsic value which outweighs its value as a tourism asset.
Its enjoyment by future generations and its long term survival must not be prejudiced
by short term considerations.

Tourism should be recognised as a positive activity with the potential to benefit the
community and the place as well as the visitor.

The relationship between tourism and environment must be managed so that the
environment is sutainable in the long-term. Tourism must not be allowed to damage
the resource, prejudice its future enjoyment or bring unacceptable impacts.

Tourism activities and developments should respect the scale, nature and character of
the place in which they are sited.

In any location, harmony must be sought between the needs of the visitor, the place
and the host community.

In a dynamic world some change is inevitable and change can often be beneficial.
Adaptation to change, however, should not be at the expense of any of these
principles.

The tourism industry, local authorities and environmental agencies all have a duty to
respect the above principles and to work together to achieve their practical realisation.

Source: English Tourist Board, 1991

Tourism can, of course, be highly beneficial and may produce substantial
income for an area. Indirectly, it may be the only effective means or justification for supporting financially the maintenance of valued countryside features.
The benefits of tourism include:

- bringing satisfaction and enrichment to visitors, strengthening a respect for
  natural and built heritage, and promoting an understanding and appreciation of other communities and cultures;
- supporting the maintenance and improvement of heritage, and acting as a
  catalyst for clearance of eyesores and dereliction;
- creating jobs and wealth, and diversifying narrowly-based rural economies;
  and
- improving quality of community life by supporting shops and services which
  might otherwise close.

These are entirely coincident with rural sustainability.

Nevertheless, tourism may have disadvantages which make it unsustainable.
These are typically associated with overcrowding, traffic congestion, wear and
tear, inappropriate development, and conflicts with the local community.
Various practical and policy measures can be taken to counteract adverse
trends. These include assessing the inherent capacity of sites to accept visitor
pressure, managing traffic, appropriate marketing and pricing strategies,
physical measures to achieve sympathetic design and site use, and involvement
of the local community (English Tourist Board, 1991). Some principles for
sustainable tourism are set out in Table 3.7

More recently, the House of Commons Environment Committee (1995) has proposed a principle of 'Best Available Place' as a basis for sustainability planning in relation to leisure and tourism. This suggests that facilities should only be provided if they fulfil a genuine need, and are as near as possible to where the main users live. Also, they should use derelict land where it is available, and elsewhere take land of the least ecological and scenic value. Careful consideration of the special qualities of a site may entail conducting an environmental assessment.

Conclusion

The idea of a sustainable locality is multi-faceted and only partially soluble. Neither settlements nor regions are likely to be set the target of sustainable development in the narrow sense of self-sufficiency. Sustainability can more realistically be related to some measure of the ecological footprint, and the ways in which this can be reduced in size and intensity. There is no rapid solution to this, nor any clear agreement over the optimum future patterns of production or development. However, the framework of flows-areas-participants may serve as an organising principle. The use of flows, or chains, will need to become more efficient; places will need to be planned and managed for optimum quality of life; and people will need to become willing and active participants in the transition to sustainability. The options for each of these will be distinctive in each region, settlement and landscape, and the quest for sustainable localities will require informed debate and much imagination.

4

Methods for approaching local sustainability

Introduction

One of the characteristics of a professional or scholarly area is the existence of a suite of distinctive methods of investigation and analysis. The production of programmes and plans for local sustainability, which is presently a rather ill-defined subject, is thus likely increasingly to be characterised by methods which give more formal expression to vague concepts. Some of these methods will be scientific and quantitative, and will thus help to sharpen up the type of data we collect and the ways in which information is reported to particular audiences. Yet this cannot be the whole story, as the scientific discourse on the environment is so complex that citizens, land managers and businesses may be deterred by its sheer technicality. The identification of environmental 'problems' and 'futures' is also socially constructed. Thus, we cannot simply look towards scientific decision-support methods. Techniques for devising local sustainability strategies will need to address not only the rational, managerial trade-offs between particular options, but also the cultures and perceptions of the people who are affected by those trade-offs.

Very broadly, we can divide methods into those which provide firm (or apparently firm) information about the environment and use it as the basis for deriving policy options, and those which facilitate citizen involvement in debates and decisions. Thus, whilst some methods will relate mainly to environmental data and trends, others will refer to partnership, awareness and participation. Although the two increasingly overlap, we may think of them as 'decision-support' and 'process-aiding' techniques respectively (Table 4.1). Both approaches are presently at rather early stages in their development, and will rely on further basic research taking place in the natural and social sciences. Perhaps the best developed of these, by virtue of its statutory status, is environmental assessment, and this is covered in Chapter 7.

Table 4.1   Some techniques for local sustainability

Examples of decision-support methods	Examples of process-aiding methods
state of environment reports; internal audits of environmental performance; development of quantitative indicators of sustainability; and environmental assessment	means of engaging public participation and debate; creation of networks and partnerships for environmental action; focus groups; freedom and facilitation of access to public registers and other environmental information; and consensus-building, conflict resolution, envisioning/future search

The importance of civic science

The complexity of environmental problems and the discussions surrounding their solution makes sustainable development an intensely political (though not necessarily party political) issue. Politics is concerned with debate, policy choices, power relations and the reconciliation of expert and popular viewpoints. Debate is central to the issue of environmental management, as there is much uncertainty about the evidence and explanations. Scientific evidence on the environment is highly contested, and much of the most fundamental evidence that we need is intractable to measure reliably. This is because environmental science is literally 'science in the environment' and presents huge problems of logistics, measurement and experimental control which are rarely encountered in 'laboratory' sciences. Also, changes in environmental conditions (such as temperatures, soil erosion, weather patterns) occur quite naturally, and it is difficult reliably to ascribe some changes to human impacts and others to natural trends. The period of the twentieth century is utterly inconsequential in geological terms, and it is highly problematic to decide whether changes which are presently occurring are the result of human pressures or whether they are part of much longer-term cycles. Thus, because there is so much controversy associated with evidence for human-induced environmental change, environmental management is associated with what sociologists call a 'disorganising discourse'.

Moreover, power relations are pivotal to our understanding of sustainable development. Sustainability is not so much a scientific issue as one of the relative power between developed and developing nations, men and women, and human and non-human species. Science can inform the decisions we take, but the decisions are unlikely to be practical and stable unless they are comprehensible and acceptable to the 'disempowered'. A particular concern here is with the relationship between experts and laity, and the ways in which lay people can at least be enabled to contribute effectively to the debate, and perhaps even take responsibility for defining and responding to major parts of it.

The informed involvement of lay people entails an adaptive approach to learning and listening by both laity and experts. Both must be willing to learn from each other's knowledge, and the 'disempowered' must be able to become significant stakeholders in decisions about policies, plans and programmes. For example, in developed countries lay people are often deeply concerned about the rate of growth of their communities as well as the levels of environmental amenity which succeeding generations will be able to enjoy. In developing countries, lay people often have the best knowledge of the types of agriculture, forestry and intermediate technology which will be sustainable in their locality. One solution is to create a 'civic science' (Lee, 1993; O'Riordan, 1994) in which complex issues, normally negotiated between experts and politicians, are instead managed through a participatory process. A true participation exercise is open to – that is, it does not feel threatened by – learning from its past errors and to giving credit, where due, to success. This may involve structural adjustments in the management of science as well as in the relationship between science policy and political decision making.

Traditionally, the way in which 'enlightened' environmental management has been approached is through multi- or inter-disciplinary approaches. These, however, are insufficient, as they are still expert-dominated. Rather, it is the bargaining process which determines possible solutions, and this is influenced by political context, the relative power of the players and the negotiative process. It could also be argued that, as this is inevitably a high-risk process, it cannot effectively be simulated, but must be acted out and treated as a social learning frame. Its essential requirements have been defined as: patience, determination, optimism, respect and commitment to a process which is open ended. O'Riordan suggests that:

> . . . civic science needs its application in the real world . . . . Its success will come through the tough bargaining that will now be a feature of the post-Rio world, as the sustainability transition is properly addressed with all of its enormous implications for the current distribution of power and wealth, and the emergence of a global citizenry.

> (O'Riordan, 1994)

One aspect of this is that citizens will need access to sound environmental information and reports, and will need to be actively incorporated into networks of influence. A further implication is that we need to be clear about the role which decision-support methods will play in relation to scientific information and partnership operation, bearing in mind that there will be overlapping expert and lay involvement in both.

Decision-support methods

Information provision
Over many years, there has been an enormous quantity of environmental information published, much of it in the public domain. Unfortunately, it has

often remained unknown and incomprehensible to the affected public, and only a sample of it has been received and interpreted by voluntary environmental groups. If we are to move towards a sustainable future, it is important that information is made available on as consistent, open and comprehensible a basis as possible. A very widespread approach has been to compile 'audits' of environmental conditions or practices.

One role for audits is to monitor and report on conditions in the wider environment (other types of organisational audit will be considered later). These are generally referred to as 'state of environment reports' (SoERs) and, over the past few years, their production has become very fashionable. State of environment reports assemble and consolidate the various categories of information, from global satellite data to local registers, which are available to professionals and laity. Their information has rarely been collected for the purpose of compiling the SoER: generally, datasets have been assembled for a whole range of dissimilar purposes, measured in different units, presented for different audiences, and collected over different 'timelines' (intervals of recording).

State of environment reports have been prepared by a very wide range of organisations. Generally, these organisations are in the public sector, as their coverage is broad rather than tied to the interests of a particular organisation. They include a spectrum of baseline information on the biophysical environment, land use patterns and trends in use of and discharges to the environment. State of environment reports generally address four key questions:

- What is happening in the environment?
- Why is it happening?
- Why is it significant?
- What are we doing about it?

They are often organised according to the framework of 'conditions, processes and responses' identified in Chapter 1 (and see below).

The purpose of SoERs is thus to provide a comprehensive and, as far as possible, objective analysis and interpretation of data which will identify significant conditions and trends in the environment. They are intended to provide a résumé of sifted information, which is chosen to be as reliable and representative as possible without inundating readers with volumes of data couched in technical jargon. All such reports must clearly be treated very cautiously as their information, whilst apparently reliable and comprehensive, is gappy, selective and has been gathered for a multiplicity of purposes, in a non-standardised manner at differing intervals. Nevertheless, interpreted intelligently, SoERs provide a wealth of information which provides a foundation for more comprehensive and systematic data collection.

The range of information to be collected for the SoER varies according to the scale of the exercise. SoERs range from district and regional, to national and international, and even global scales. Their content varies according to purpose, and to availability of datasets and human resources at the disposal of the responsible agency. Ideally, depending on the budget, a SoER should be a team effort involving public officials, universities, private consulting firms, environmental organisations, industries and other user groups. In practice, it is often a more parsimonious, in-house exercise, but one which may be valuable nonetheless. The range of information included in a SoER has been strongly influenced by the framework set in a report by the OECD (1991), although it is invariably adapted to suit local priorities and availability of data. The main contents of a SoER typically centre upon:

- air;
- water;
- waste;
- noise;
- energy;
- land and agriculture;
- wildlife;
- landscape and townscape;
- open space;
- transport;
- coast;
- population issues; and
- minerals.

The most comprehensive exercise to date in Europe has been that of the Dobriš assessment (Stanners and Bourdeau, 1995), which reported on key environmental problems and provided an analysis of the strengths and weaknesses of datasets associated with each indicator.

During the early stages of environmental reporting, the emphasis has been on assembling disparate information and drawing it together into as user-friendly and consistent a format as possible. However, as the datasets have become better organised and some of their more serious omissions corrected, the trend has been towards summarising them in numerical and visual ways so that they can more readily be interpreted by users. The main numerical approach has been to derive quantitative indicators, which give a clearer impression of the ways in which changes are taking place, and the effects which policies are having. A greater sense of association and significance may be achieved by integrating these indicators on a spatial or regional basis so that they can be mapped for areas – or 'ecozones' – with which citizens can more readily identify.

Whilst numerical indicators are valuable as formal reporting devices about particular performance topics, it has been suggested that more forceful effect can be obtained by presenting information in an integrated manner. The types of themes around which information may be organised comprise (International Institute for Environment and Development, 1995):

1. global limits – ideas about the world's carrying capacity, in terms of environmental resources and sinks for pollution and wastes;
2. fair shares – reflecting the notion that each inhabitant has an equal right to the earth's limited carrying capacity;
3. meeting needs – assuming that food security concerns should take precedence over the allocation of land resources to 'luxury' products;
4. sustainable trade – reflecting a type of international trade which is based on ecological surpluses and minimisation of transport; and

5. full life-cycle costs – internalising full social and environmental costs at each stage of the life cycle from extraction through production and consumption to disposal.

These contain some radical ideas which politicians (and citizens) often find threatening, especially as the ability of these methods accurately to reflect the issue in question has been seriously criticized. Consequently, their use remains controversial, though their potential value as powerful, integrative mechanisms is now widely acknowledged.

Environmental indicators are comparable to yardsticks developed for financial and economic purposes, such as the retail price index, gross domestic product and stock exchange index. Local state of environment indicators are typically of two kinds:

- those which reflect the quality of the environment, the stresses on it and the management responses to both quality and stress; and
- those which reveal the environmental efficiency and impact of key sectors, such as agriculture, energy, transportation and industry.

Gradually, a third category of indicator is emerging which conveys the success of the 'process' used to inculcate sustainable behaviour within target communities. Environmental indicators may attempt to describe: local environmental quality; local environmental performance (e.g. levels of economic activity, public opinion, number of protected areas); or environmental 'accounting' (integration of environmental concerns in economic policies). This last approach is mainly relevant to national, rather than local, audits and it seeks to integrate economic and environmental elements, reflecting the value of natural resources and costs of pollution in national accounts such as the GDP. It has, however, also been used at the regional scale.

It is now fairly common practice to follow the tripartite OECD classification of indicators into those which measure the *condition* of the environment, those which depict the *stress* imposed on the environment by human activity, and those which refer to our *management* of (or 'response to') those stresses. The relationships between condition, stress and response is shown in Figure 4.1. Some exponents of the indicator approach have, as previously noted, argued strongly for the production of a single index of environmental quality, so that it can be reported on a succinct and regular basis. Given the host of environmental issues, and of agencies with particular environmental interests, this has not proved possible. However, recent research in the UK has suggested that it would be desirable to report widely on a limited selection of 'headline' indicators, whilst a range of more detailed measures could be collected for particular interest groups. The initial set of indicators for the UK, proposed by an inter-departmental Working Group, was published early in 1996, mainly with the intention of stimulating debate about the meaning of sustainable development and how we measure progress towards it (HMSO, 1996). An actual example of local state of environment indicators is given in Table 4.2.

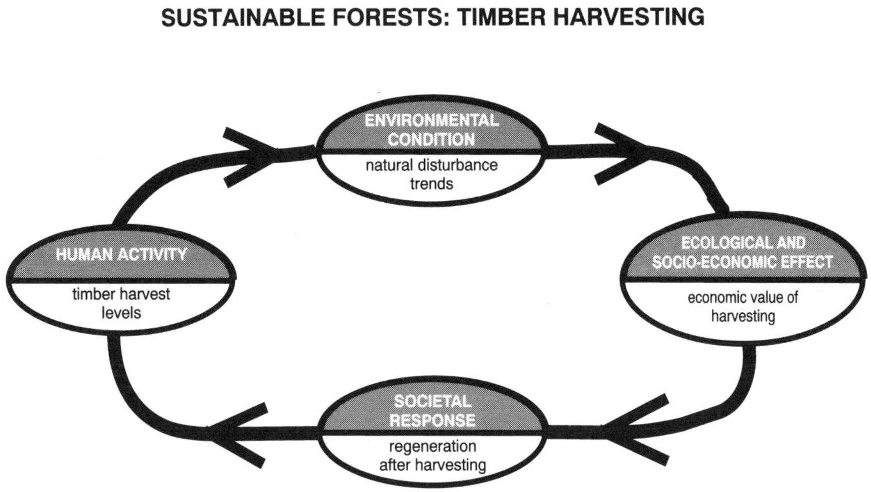

SUSTAINABLE FORESTS: TIMBER HARVESTING

Figure 4.1 Example of linked indicators in State of Environment reporting

Source: based on Environment Canada, 1995

As indicators become more widely used, so their weaknesses and limitations become more acutely exposed. Apart from the inherent uncertainties contained in collecting and interpreting environmental information, problems associated with summary indicators have mainly been related to their unsystematic nature and lack of fitness to social and policy objectives. Brugman (1994) has also criticised their 'join the dot' mentality, and argues that they require:

- constant observation, rather than periodic measuring;
- careful analysis, rather than common denominators; and
- regular course correction, rather than annual claims of progress.

Ideally, a wider debate should be conducted on the identification of indicators amongst those – including the public – who are likely to use them.

The following account describes an idealised approach, and is based broadly on the steps taken during the preparation of federal SoERs in Canada, where there is perhaps the longest pedigree of this kind of work. A first step is to identify the goals held by society which relate to environmental quality and responsibility, and then to seek indicators which relate closely to them. Without this clear connection to actual, socially defined priorities, the SoE Report is likely to lack trust and confidence. Next, a conceptual framework is needed to set the bounds and structure of the investigation, and to form a practical basis of operation. There must be a clear rationale for reconciling the inherent complexity and inter-connectedness of the environment with the need for simplicity and efficiency of information presentation. Beyond this, the team

Table 4.2   Example of State of Environment indicators proposed for use by Lancashire County Council

Topic area	Indicator (examples only)
more efficient use of resources and less waste	domestic waste production per person, and commercial waste production per unit of GDP, per year percentage of domestic waste recycled, by type, per year domestic energy consumption (gas and electricity) per person
lower levels of pollution	nitrogen dioxide levels in air percentage and length of rivers and canals in 'good', 'fair', 'poor' and 'bad' quality categories number of bathing waters passing European standards, and percentage compliance of bacteria samples taken from water
a more diverse natural environment	area of protected wildlife habitat and area damaged or destroyed change in populations of selected, key, threatened species that are representative of the area
basic needs for everyone which are met locally	percentage of the population within walking distance of basic services areas of prosperity and deprivation percentage of drinking water samples that fail European standards for lead
more opportunities for work in a diverse economy	percentage of people living below the poverty line numbers and percentage of people unemployed for more than one year
improvements in health	mortality rates for a range of diseases compared to the national average percentage of the population diagnosed to be asthmatic
access to facilities, goods, services and people whilst protecting the environment	percentage of population travelling to work by car, bus, train, walking, cycling and other mode percentage of people satisfied with bus and rail services
less fear of crime and persecution	percentage of people fearing violent crime and burglary by gender and age compared with total numbers of reported crimes per 1,000 population in each category percentage of people believing racism is a local problem compared with the number of reported racial incidents where further investigative action is taken

access to education, training and information	number of adults enrolled on part time or evening education and training courses percentage of pupils with specific levels of attainment in state examinations
people having a say in decision-making	percentage of electorate voting in local elections number of self-help, community and campaign groups in the county
people valuing the neighbourhoods and communities in which they live	area of open space available for community use per 1,000 people percentage of local citizens who volunteer at least 50 hours of their time to civic, community or non-profit activities

Source: LCC, 1995

must generate some selection criteria against which to judge potential indicators: whilst these criteria may often be somewhat subjective and based on value judgements, they should nevertheless prevent uncritical adoption of inadequate or misleading indicators.

According to the Canadian practitioners, an environmental indicator should, for inclusion into an SoER, as a minimum:

- be sensitive to changes in the environment;
- be easily accessible to the auditors, or be readily measurable;
- be cost effective in terms of data needs and monitoring requirements; and
- have identifiable thresholds or limits which indicate where significant changes are taking place.

In a different context, Hams et al (1994) and Barton and Bruder (1995) have also proposed more challenging set of conditions, and these are summarised in Table 4.3. Few indicators will immediately satisfy all of these requirements, and important issues should not be jettisoned from a SoER because, for example, they require additional research to clarify critical thresholds.

Table 4.3 Principles for selecting environmental indicators

Environmental indicators should cover:

- environment and economy;
- environment and health;
- environmental rights, equity and quality of life;
- impacts of lifestyles and consumption patterns; and
- environment and culture.

They should:
- provide a representative picture of conditions, processes and society's responses;
- be simple, easy to interpret and show trends over time;
- be responsive to changes;
- provide a basis for international comparisons;

- include a target or threshold so users can assess the significance of changes;
- be well founded in technical and scientific terms;
- be readily available; and
- be capable of being updated.

They should conform to a set of general criteria:

- Scientific validity – are they replicable; do they accurately reflect actual phenomena?
- Availability of longitudinal datasets – do records exist over a sufficient period of time and at suitable intervals to enable tracking of long-term trends?
- Responsiveness to environmental change – if conditions improve or worsen, will the indicator rapidly respond to this change?
- Relevance to social objectives – does the indicator relate to things which society deems important?
- Sensitivity to thresholds – if there is a critical threshold (e.g. a legal limit of pollution levels, or a limit at which an environmental system starts rapidly to deteriorate), will this be picked up by the indicator?

Source: after Hams et al, 1994; Barton and Bruder, 1995

Using criteria such as these, a durable suite of indicators may then be selected, often following the condition-stress-response model (Table 4.4) It is then typical for the team to consult with data holders, experts and potential users. Consultations with subject-area specialists can help refine the preliminary choice of indicators, whilst continuing liaison will help organisations to augment the data which they currently collect so that it will dovetail with longer-term intentions about SoERs. Finally, once the indicators are in operation, they need to be validated (i.e. checked for factual accuracy)and evaluated (i.e. judged in terms of value for money and fitness for purpose). In particular, the team must be able to confirm that the indicators are communicating effectively the message to the intended audiences.

State of Environment Reports are most commonly available in published format, and have proved popular with a wide variety of users. However, increasing attention is being given to the use of electronic media, especially CD-ROM (read only and interactive) and geographic information systems (GIS). One of the pioneering local authorities in British SoE Reporting has been Lancashire County Council, where use of GIS has been central to the environmental auditing process, both in producing the initial Report and in subsequent updating exercises. Information is stored on an Arcinfo GIS in the Planning Department, whilst the more user-friendly ArcView facility has been used to create a network of environmental information centres throughout the county, and to inform schools and colleges. In County Durham, SoE information has been transferred from the ArcView system and converted into a form which can be read by the types of computers used in local schools. A package is also being used to set up environmental information centres in libraries and other community facilities (Masser and Pritchard, 1994). Advances in Canada since the production of the second Federal State of Environment Report in

Table 4.4 Framework for identifying a suite of environmental indicators

Categories of environmental resources	Measures of environmental condition (exposure/response)	Measures of human activity stresses	Measures of management responses
present ecosystem state (air, water, biota, land)			
environment-related human health measures (e.g. drinking water, waste mangement)			
natural economic resources (e.g. energy, forestry)			

Source: Government of Canada 1991

1991 have been to transfer environmental indicator bulletins to Internet, and produce a CD-interactive for use in schools and homes (Selman, 1994a) culminating in the multi-media format of the 1996 report (Government of Canada, 1996).

As a rejoinder to electronic environmental information reporting, it should be recognised that these systems can convey a false impression of accuracy. There is a tendency to believe that, because information has been downloaded from a real-time electronic system, it is up-to-date, reliable and definitive. As has been noted, SoE information is unreliable for a host of reasons, and has to be interpreted cautiously and intelligently. Hopefully, civic science and active environmental citizenship will lead to an informed and critical awareness of both the value and limitations of environmental indicators.

Capacities and thresholds
An enduring concept in environmental management has been that of carrying capacity, defined ecologically as the maximum population of a particular species that a given habitat can support over a given period of time. When the carrying capacity is exceeded, if the species is not able to colonise new areas, the habitat may become irreversibly degraded. The organism's food supplies may become over-exploited to the point of incipient collapse, and the organism itself displays increased morbidity or mortality. The analogies with human settlements and use of the environment have a distinct resonance. However, human carrying capacity, or the critical threshold beyond which living conditions become intolerable or a particular environmental amenity is over-exploited – whether at global, site or intermediate level – is virtually impossible to specify. The threshold is very elastic, and varies in relation to the

sophistication of technology, cultural expectations about material conditions, or specific planning and management objectives.

Although previous attempts to define capacities have not been notably successful, even in relatively simple contexts such as recreation, the concept is enduring. Recent studies have interpreted environmental capacity as the capacity of an area to sustain a particular activity or level of use. The object is to try and identify the level at which unacceptable or irreversible damage is likely to occur, and to ensure that such change does not happen. A key notion here is the 'limit of acceptable change' (LAC) which can be used to identify indicative capacities and requirements. A caveat is that capacity planning should not imply that sites should be planned up to capacity all the time, and that the limit of acceptable change is not the same as the optimum.

Limits, though, may be actual or notional. They may be simple, two-dimensional spatial capacities, such as the remaining stock of 'undeveloped coast'. Alternatively, they may refer to a more fundamental concept of stability, such as the exploitation rates of renewable resources or extraction rates of non-renewable resources. Equally, they may be more culturally defined in terms of legal and political capacities, including the capacity of people to accept environmental losses and gains (c.f. Pritchard, 1994). As there is no simple definition of capacity, the concept cannot be reduced to a formal technique for the evaluation of alternative planning proposals: rather, a deeper respect for environmental systems needs to permeate the production of development policies. Pritchard has thus argued that development aspirations should be fitted around environmental constraints, rather than the other way around. Capacities and thresholds can then be translated into targets and objectives, with local ones expressed in terms of their contribution to regional capacities, and regional ones in terms of national capacities. If scientifically or politically defined thresholds are not available then, it is suggested, policy makers should adopt a precautionary approach.

It is helpful to define several different types of capacity in urban and rural environments (Table 4.5), and to be aware of the relative ease of measuring them. All of these are open to considerable flexibility in their definition, interpretation and management. One view is that the concept of capacity is relatively straightforward when applied to factors such as the existing building stock where the stock is reasonably stable and exogenous variables are relatively predictable. However, it has less validity when applied to elements such as natural habitats that have innate regenerative abilities, and where policy decisions, such as altered management regimes, can easily alter the context. In many situations, it may be more helpful to refer to specific thresholds or levels of activity which have been defined in relation to the need for policy review, rather than in terms of specific biophysical changes. This reflects the fact that the definition of thresholds, whilst informed by technical analysis, is essentially based on political judgement, and must be debated in terms of their trade-offs rather than treated as neutral facts.

Table 4.5   Types of environmental capacity

built environment	capacity of building stock
	capacity of infrastructure
natural environment	capacity of wildlife habitats/landscapes
	capacity of the natural resource base
human environment	capacity in relation to traffic
	capacity in relation to social impact

Source: based on DoE, 1993b

In the urban context, capacity has most successfully been applied to historic towns, which face a variety of traffic, development and tourist pressures. A pioneering application to the city of Chester entailed application of a two-stage methodology, namely, identification of a capacity framework, and production of guidelines. The first of these set out to identify and understand the special historic qualities of the town, which can be thought of as 'critical environmental capital'. This comprises both physical attributes and the activities or demands on a place. The second entailed identification of tensions in order to appreciate the kinds of pressures which the current scale and mix of human activities were having on the town. These included traffic congestion, conflicts of need between visitors and residents, and the desire to provide for new business, housing and car-based retailing without detracting from the town's special qualities.

In order to focus on key issues and indicators, a series of technical studies were conducted into factors such as the incidence of road accidents, rush-hour transportation capacity, and changes over time to the urban fabric of sensitive areas. These were complemented by a series of perception studies with local residents and organisations based on group and panel meetings, and supplemented by social surveys. Indicators were also used to help measure change over time, to assess whether an issue was becoming more serious or not, and the likely impact on the issue of a change in circumstances. One of the most revealing indicators in the Chester study was the change in the proportion of 'solid' and 'void' of the urban grain, with a 60% solid and 40% void emerging as a critical threshold. The sum total of all the special qualities, the indicators of change and the possible critical points of future change were used to make up the capacity framework (Figure 4.2).

The production of guidelines enabled the team to test how the capacity framework might be affected by different types and levels of activity. Long-term scenarios, reflecting the effects of different mixes of activities based on broad land use strategies, were used to demonstrate the different ways in which the city could function as a whole. An integral part of the exercise was to identify the impacts of a scenario both on the city core and the whole city, and to assess the implications of these for key issues and the capacity framework. In this process, it is useful to be able to compare the experience of other historic cities in accommodating particular types of change, and so

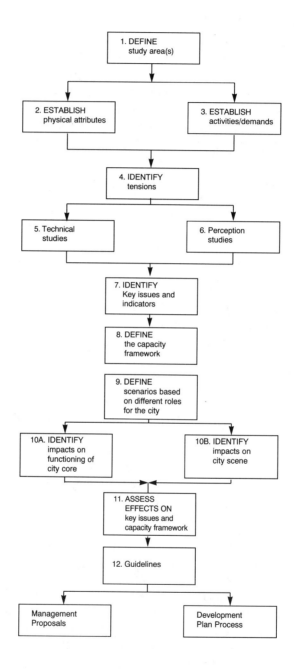

Figure 4.2   Researching the capacity of an historic city: steps 1–8 refer to the identification of the capacity framework, and steps 9 onwards, to the production of planning and management guidelines.

Source: Roebuck and Gurney, 1995

Table 4.6   Benefits from applying the capacity methodology to a 'real life' example

- definition of the critical environmental capital of the city under study – the urban edge, compact nature of the city, key water and green spaces, historic buildings and monuments, archaeology and shopping facilities;
- definition of ten key issues and sixteen environmental indicators to assess the impact of development and capacity thresholds;
- guidelines for future development planning, ranging from broad directions to specific observations on important skylines, views and streets;
- long-term view of the capacity of the city;
- landscape and open space appraisal and framework;
- extensive use of perception studies and techniques amongst various groups to define what makes the city special;
- a review of various development scenarios and their impacts on the capacity framework; and
- public consultation on the study's findings.

Source: Roebuck and Gurney, 1995

comparative indicators were developed to help identify where and when critical thresholds may exist. The guidelines eventually led to the production of a final report, which presented environmental information in terms compatible with normal planning exercises, offering planners the opportunity to look further ahead than is normally possible in the development plan process. The team identified a number of benefits from applying the methodology to real life (Table 4.6).

The notion of capacity in rural areas has been explored over a longer period of time, especially in relation to tourism and recreation. In relation to tourism, it can be assumed that every site or visitor destination has a threshold or capacity beyond which the numbers of visitors begin to cause serious damage to the resource itself, or result in a significant deterioration in the quality of the visitor experience (ETB, 1991). This capacity varies enormously between different sites, and is also influenced by visitors' expectations regarding crowding or solitude. Various concepts of capacity are, however, useful in a wide range of situations, and these include:

- the level above which the ambience and character of the place is damaged and the quality of the experience threatened;
- the level above which physical damage occurs;
- the level above which irreversible damage takes place; and
- the level above which the local community suffers unacceptable side effects.

Capacity can also vary according to weather and seasonal conditions, as well as the level of investment in site facilities. Determination of capacity is thus not an exact science, and involves a good deal of value judgement on the part of the site manager. Usage of tourist and recreational sites can be constrained by a variety of approaches ranging from prevention of access to more sophisticated techniques for limiting and controlling visitor flows.

An application of the LAC approach to the rural environment is provided by the management of the Cairngorm skiing area in Scotland. Here, the LAC concept provided a framework for identifying key issues and attributes to be monitored, setting quality standards and listing the management responses that should be followed if the standards are not met. The approach was considered to be particularly well suited to complex land-management circumstances since the decisions were taken by a representative group, rather than by the commercial site operator alone. The first stage involved compiling an inventory of resources to be managed which, fortuitously, was available from a previously conducted environmental impact statement, complemented by some additional survey work. This inventory helped the monitoring group identify the main environmental issues and a list of key environmental attributes; these latter could be monitored according to a specified set of procedures. The final start-up stage was to set preliminary LAC values for the key attributes, and the management responses to follow non-compliance. Once established, the system operated on an annual cycle with the LACs gradually becoming more sophisticated as experience grew (Bayfield and McGowan, 1995) (Table 4.7).

Table 4.7 An application of the 'limits to acceptable change' model in environmental management

Attribute	LAC 1988	LAC 1994	Management response
path widths	doubling 1987 widths	seven individual path widths (1.5–3m)	rebuild paths or restrict use
bare-patch size	4 sq.m or 4 linear metres of freshly exposed bare ground	unchanged	seed or turf
vegetation damage along tows	90% vegetation cover	90% original vegetation cover	seed and fertilise
damage on summit ridge	10% increase in bare ground	20% increase in dead moss	restrict access
success of re-instated vegetation	60% cover after one year	abandoned	repeat re-instatement
queuing times	15 minutes	10 minutes	redistribute skiers
litter	25 items/km	15 items/km	increase litter collection
fires	one	abandoned	review preventive measures
wear on named mountain sites		extent in 1994	prepare landscaping proposals
erosion on named tow ditches		depths in 1994	reconstruct eroding sections

Source: Bayfield and McGowan, 1995

The idea that there are environmental thresholds which can be overcome only at the expense of serious and irreversible damage to the natural environmnet is also expressed in the ultimate environmental threshold (UET) method (Koslowski and Hill, 1993). This technique has close affinities with methods of threshold analysis which have been widely used in urban planning. In essence, whereas its urban counterpart identifies an economically viable 'solution space' (i.e. satisfying policy constraints and giving a positive return on capital), in which a unique optimum solution may be sought, the UET approach seeks an ecologically sound solution space within which development proposals must be generated and contained. Four types of threshold are identified, namely:

- territorial (spatial) – the area over which a given activity can take place;
- quantitative – the level up to which an activity can be developed;
- qualitative – retention of factors important to environmental quality; and
- temporal – the acceptable development rate, or time periods in which development can take place.

Elements of the environment can be expressed in degrees of 'uniqueness', 'transformation' (extent of change from pristine or sustainable condition) and 'resistance'. Environmental thresholds are defined as those imposed directly by natural resources, and they represent significant and specific development limitations. Some of them also indicate 'final' limits of what the natural environment can take without being irreversibly damaged or even destroyed, and these are the 'ultimate environmental thresholds'. Although the method has yielded only limited results in practice, its authors have produced varied case study evidence from national parks, tourist developments and fisheries.

Corporate environmental reporting

State-of-environment reporting is an attempt to set out, on an objective and comprehensive basis, environmental conditions in a given geographical area. At the level of the individual organisation or enterprise, the task is essentially that of auditing corporate environmental performance. Auditing is the act of verifying impartially the performance of an organisation relative to pre-set standards or targets. Traditionally, it has been used in industry and government as a means of assuring financial probity and evaluating corporate performance more generally, but during the last thirty or forty years it has increasingly been extended to non-commercial applications. The first departure was into 'social auditing' through which public sector agencies and some businesses could demonstrate that their activities had some desirable wider benefits beyond cost-efficiency or shareholder profits. Much more recently, auditing has been extended to the evaluation of environmental performance, in two ways. First, as previously discussed, it has been used to record changes in the condition of the environment, relative to local, national and international indicators of ambient quality. These may be thought of as 'external' audits. Second, it can relate to the systematic and objective assessment of a public or private sector organisation's policies and practices, as they impinge on the environment. Since these refer to the way in which practices operate inside an organisation, they are known as 'internal audits'. Internal environmental audits are introduced here, and are discussed more fully in relation to local government in the next chapter.

The EU sees environmental auditing as 'a management tool comprising a systematic, documented, periodic and objective evaluation of how well environmental organisations, management and equipment are performing, with the aim of contributing to safeguard the environment by facilitating management control of environmental practices and assessing compliance with company policies, which would include meeting regulatory requirements and standards applicable' (Commission for European Communities, 1993). The basic procedure for internal auditing is borrowed from general quality assurance systems, namely, a culture of continuous improvement supported by explicit procedures. These procedures typically comprise:

- contract review – ensuring that contracts are sufficiently clearly specified before commencing work;
- quality plans – each contract is associated with a plan which includes the project brief and the resources to be used;
- design control – detailed procedures and calculations for the work to be undertaken;
- document control – control of all documents and data relevant to the contract;
- quality audit – internal auditing to verify whether quality activities are being performed as planned, and corrective action taken if not;
- training – appropriate staff training; and
- other procedures – purchasing, servicing, inspection, etc.

This approach differs from 'quality control' – a quality standard for a product – in that the entire system is under scrutiny. The core of *environmental* quality assurance is an environmental management system (EMS) which must be adopted by an organisation in order to ensure a cycle of continuous improvement in relation to the organisation's environmental performance objectives (see CEC, 1993; BSI, 1987, 1992). The EU Regulation defines an EMS as:

> . . . that part of the overall management system which includes the organisational structure, responsibilities, practices, procedures, processes and resources for determining and implementing environmental policy.

A very closely related activity is the 'environmental review', an example of which is provided by O'Brien and Gibbins (1994) in respect of Newcastle airport. Their approach to the initial stage of the review was to conduct a SWOT analysis (i.e. an examination of the strengths, weaknesses, opportunities and threats faced by the operator), based on a short questionnaire (Table 4.8). This was accompanied by a desktop review, to assess the current position of the business with regard to the environment, that is, its baseline

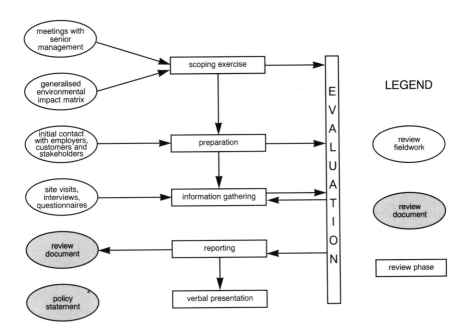

Figure 4.5   Schematic representation of the initial environmental review process
Source: O'Brien and Gibbins, 1994

environmental performance. Next, an environmental policy for the operator was developed, representing a formal statement which defined the organisation's objectives and strategy for the environment. This was derived from the baseline analysis and desktop review. The policy has to steer a middle path which is broad enough to cover the whole organisation and its sphere of influence, yet which also addresses the specific direction necessary to achieve environmental excellence. The policy serves both as a practical working document and the organisation's 'public face', so that it must be an effective communication document, not shrouded in technical jargon. An EMS is then used to provide a framework within which audits and reviews are conducted, and is integrated with the overall business management activity. Finally, environmental audits and reviews are periodically conducted to assess the organisation's success in meeting its policy objectives and the continued effectiveness and relevance of its policy and EMS (Figure 4.5).

Corporate environmental reporting is, nowadays, not an easy option, nor a short-cut for an organisation wanting to give itself a gloss of public respectability. In the context of local authority environmental quality auditing,

Table 4.8   An approach to a short questionnaire audit for environmental review

---

A series of two page questionnaires was used for:

airport company employees

middle management (covering their personal environmental concerns, environmental effects associated with their particular company role)

site operators (covering whether their organisation had an independent environmental policy in operation, their views on the possibility of a site-wide policy)

Source: based on O'Brien and Gibbins, 1994

Barton and Bruder (1995) point to the significant costs, as well as benefits, accruing from the process. Costs include a substantial commitment of re-sources. These may include the possible disruption of normal work, extra costs of any actions recommended by the audit, an increased risk of public disap-pointment, and the publicisation of embarrassing actions or inactions. Set against these, the benefits include 'heroic' ones of genuine major improvements in environmental quality and increased public environmental awareness, and 'humdrum' ones of legislative compliance and expenditure savings. They con-sider that auditing can facilitate the 'networking' of diverse decision areas and the implementation of specific internal actions. The latter comprise:

- staff training;
- transport audit;
- waste audit;
- energy audit;
- eco-management and audit coordination;
- linkage to compulsory competitive tendering;
- progression towards registration (with BS7750 or EU eco-management scheme);
- environmental budget appraisal (e.g. developing a system of considering budgets in terms of environmental value for money); and
- contribution to Local Agenda 21.

Barton and Bruder also note four significant future prospects. First, they iden-tify 'implementation mechanisms', through which verbal support is turned into financial backing, and where local authority-wide, and general public, support is obtained. Second, they call for high level local authority support for the effective inclusion of the lay public in environmental programmes. Third, they advocate the coordinatory role of the local authority, at the centre of an overlapping system of environmental management; without this, problems are merely likely to be shifted rather than solved. Fourth, they argue that corporate environmental reporting can assist the definition of sustainable development 'in practical terms'.

Building visions and resolving conflicts

Quantitative evaluations of environmental performance are valuable contributors to the decision-making process, but they may have the unintentional effect of disconcerting lay members and deterring public participation. Indeed, historically, environmental planners and managers have seen people almost as a complication. The laity have been perceived as obstructing optimal solutions with their own prejudices and vested interests. The very essence of environmental issues, however, is that they are 'wicked' problems, in which equally legitimate interests will always be in conflict, and which can only be resolved (if at all) through an explicit process of bargaining between affected people. Consequently, methods and techniques for local sustainability must embrace those which are primarily concerned with creating visions, resolving conflicts and building consensus. Considerable investment may be necessary to achieve these aims, but there is a long-term pay-off in producing a more informed citizenry and securing genuinely sustainable policies and projects.

The notion of 'envisioning' – enabling lay people, along with technicians and policy makers to anticipate environmental change and thereby contribute to its management – is inherently appealing but very difficult to realise. Lay people may have negligible understanding of key terms in urban design or ecological habitats which professionals find relatively easy to visualise; they may have little facility even for such basic skills as interpreting a map. Various approaches have therefore been devised to make alternative futures more real to them, so that they can engage in a purposeful dialogue with planners and policy-makers. Visions involve two basic elements: ideas and pictorial representations.

Pictorial methods involve producing displays of possible futures, and this has typically been approached in one of two ways. One technique is to provide public participants with maps, landscape photographs and/or aerial photographs of their area (as an alternative to maps), and invite them to mark perceived problems and suggested solutions on them. These are the environmental version of 'planning for real' exercises in urban settings, and are discussed more fully in Chapter 6.

An alternative approach is to engage the public in debate about environmental 'futures' which might arise from various policy options. This typically entails high quality artistic visualisations of alternative outcomes, for instance the different types of rural landscape which might emerge under different types of farming and different levels of public policy support. If acceptably high picture quality can be obtained, these impressions might be generated directly from a geographic information system: some of these now possess such sophisticated graphic and plotting facilities that their 'virtual futures' look quite realistic. Alternatively, artists' impressions may be used: in the Yorkshire Dales (O'Riordan et al, 1993), for example, visitors were shown a series of paintings of possible future landscape types, ranging from retention and restoration of the traditional drystone wall pattern of enclosure, to a relatively featureless 'modern' agricultural landscape. This was accompanied by a board game, involving players in decisions about land use investments and associated environmental consequences. In Norway, the alternative outcomes of agricultural policy options have been visualised artistically (Emmelin, 1996), and these have proved particularly helpful in drawing out some unexpected consequences. The method entails quite a complex sequence of analytical stages, which can broadly be summarised as:

- gathering baseline information on policies and the local environment;
- generating scenarios about possible future alternative appearances of the landscape;
- projecting a business-as-usual trend for the landscape, on the assumption that it would change even within a stable farming and policy framework; and
- undertaking an analysis of the nature and consequences of business-level decisions at the individual farm unit scale.

This results in a range of visual expressions of possible futures, which in practice can yield some significant challenges to our conventional expectations.

Consensus on ideas is a more abstract concept, but one which can be aided by specific exploratory techniques. These relate to areas of conflict-resolution and consensus building. Access to environmental resources has widely been associated with competition and conflict; in affluent societies where amenity is highly valued this has often led to NIMBY (not in my back yard) attitudes, where externally generated proposals are opposed by locals. Indeed, there is a considerable degree of justification for NIMBYism, as it often reflects quite legitimate concerns of imbalanced power relations between central authority and local aspirations. Other styles of environmentalist opposition, such as NOPE (not on Planet Earth), NODAM (no development after mine) and BANANA (build absolutely nothing anywhere near anyone) are far less defensible. However, the depth of dissent associated with sustainable development issues is unhelpful and unproductive, and forms part of a more general phenomenon of confrontation and dispute. Thus, environmentalists have become attracted to the practice of consensus-building as it has developed within business and industry.

The classic expression of this, *Getting to Yes* (Fisher and Ury, 1991), asserts that, by seeking common ground rather than each party digging itself into an entrenched position, it is possible to build a mutually beneficial consensus. Consensus is very different from compromise, which merely implies a 'least bad' solution capable of holding a temporary ceasefire. Consensus building is often referred to as a win-win solution and compromise as a lose-lose solution, referring to the value of the outcome for each 'side'. Although Fisher and Ury's prescriptions are quite complex, their basic premise can be expressed in four points:

- separate the people from the problem;

Table 4.9   Some factors contributing to success in environmental consensus-building

- Stakeholders involved from the start and treated equally.
- Stakeholders given strong reasons for participating.
- Independent third parties (mediators/facilitators) work with the stakeholders.
- Rational and emotional interests are recognised and valued, and focus is on people's concerns rather than their personalities.
- Stakeholders are accountable to their constituencies.
- Time-frame is sufficient for people to get to know and trust each other.
- Solution cannot be pre-determined by any one stakeholder.
- Process does not terminate with agreement of solution, but continues with a commitment to implementation and monitoring.
- Commitment to abide by outcomes.
- Openness, honesty, trust and inclusiveness.
- Shared responsibility for success, rather than assuming 'the Council should do something about it'.
- Common information base.
- Mutual education and exchange.
- Decisions made by consensus.
- Shared responsibility for outcomes and implementation.
- Flexibility in framing and implementing solutions.
- Improved horizontal co-ordination.
- Non-discriminatory practices.
- Improved availability and use of environmental feedback.

Source: based partly on Environment Council, 1995.

- focus on interests, not positions;
- invent options for mutual gain; and
- insist on using objective criteria.

Other factors contributing to the successful achievement of consensus are summarised in Table 4.9.

In sustainable development, therefore, a frequent issue is how to move from NIMBY to YIMBY (yes, in my back yard) (Massam, 1988). This, of course, does not merely imply that local people are bribed to accept a flawed option, but that an intrinsically just and sustainable solution is achieved. This may entail two broad approaches, one based on the mediation of conflict and the other on the building of consensus before a critical stage has been reached; both of these, ideally, have a skilled negotiator at their disposal. Douglas (1992) has described the process associated with the location of a hazardous waste treatment plant near Edmonton, Alberta. A waste-management corporation was formed at provincial level and proposals invited from private firms for the construction and management of the facility. A siting strategy involved a spatial search to identify environmentally unconstrained land, and extensive community involvement then entailed 120 community meetings throughout the province to explain the nature of the facility. Following this, some two-thirds of the affected local authorities requested assessments for plant suitability, although some municipalities subsequently withdrew their bids following analysis of the constraints or public opposition. When eventually narrowed down to two contenders, local plebiscites were held, both resulting in high turnouts and votes strongly in favour of the facility, and one was finally selected.

Successful though this participatory process was, it does not always lead to such positive results, and similar procedures in other more populous or sensitive amenity areas have resulted in impasse. However, despite some countries' (including the UK's) preference for traditional adversarial inquiries, others have involved techniques which incorporate skilled conflict mediation, mainly in relation to individual projects. This approach to local environmental diplomacy or negotiation is, for example, evidenced by the use of 'informal hearings' in the New South Wales courts. Broadly speaking, solutions to environmental disputes may be aided by three types of trained third parties, namely: an *arbitrator* who, in an adversarial situation, adjudicates between 'winner' and 'loser'; a *mediator*, who seeks to find common ground between opposing parties, and to identify a mutually acceptable solution; and a *facilitator*, who enables different groups to explore creative and perhaps unanticipated options for future change.

'Communicative' techniques have tended to be used in relation to specific project proposals, or perhaps policy options, but some more strategic applications are now becoming popular. In particular, 'round tables' have been used in the production of some sustainable development strategies. Although the use of this term has become blurred, its basic meaning is simple. The idea of a round table is simply one where there is no head: thus all participants, be they business people, council employees, or community-based organisations, are seen as being of equal stature. There may be a chair or facilitator, but that person is there to take the discussion forward rather than define the content or outputs (UNA, 1995). Mormont (1996) has also described the preparation of 'river contracts' and 'nature contracts' in Belgium, to secure agreement between conflicting parties about future policies and actions towards particular aspects of the local environment. Again, there is no ranking between parties, and pursuit of the contract's objectives must rest on mutual respect and moral obligation. However, in this case, the goals are assisted by the central involvement of a very prestigious national charitable foundation, which is able to establish a basis of operation for the projects, and open doors to very senior government officials. It has elsewhere been suggested that this type of venture does not always need to assume an equality between participants, but merely that they will be able to derive mutual benefit and that they will not be patronised.

Methods of consensus building hold much in common with the notion of 'participatory inquiry', which comprises a range of methods concerned with problem diagnosis and community empowerment. Although this label covers a diversity of approaches, it is unified by a set of common principles (Carew-Reid et al, 1994), notably:

- a defined methodology and systematic learning process, focused on 'cumulative learning';

- multiple perspectives, seeking diversity rather than masking the complexity of situations in terms of average values;
- group inquiry processes, which mix disciplines, sectors, professionals (outsiders) and local people (insiders);
- context-specific applications, implying a methodology which is sufficiently flexible to adapt to local circumstances;
- expert involvement mainly on the basis of 'facilitation'; and
- a commitment to action, so that the inquiry process is intended to bring about change (which itself focuses the nature of debate between participants).

Some of the many components available to 'participatory inquiry' approaches include 'team contracts', rapid report-writing, role reversals, focus groups, and maps and profiles of local activities. A further popular variant is the 'future search' method, in which small groups 'brainstorm' potential options. Careful attention is given to room layout and the inclusion of senior people within an organisation (up to and including Chief Executive/Managing Director) to ensure that ideas are not simply forgotten and disavowed once the exercise is over. Participative inquiry approaches sound attractive and have, indeed, been used to powerful effect in a number of sustainability projects. However, it is also advisable to be aware of the various practical barriers to effective participation, such as:

- their requirement for significant additional time and effort;
- the attitude and culture changes which are needed from 'professionals';
- the difficulty of measuring achievement in conventional, numerical terms;
- unwillingness to devolve significant financial resources;
- lack of commitment to continuity and long-term evolution of approaches; and
- sensitivity of political issues (e.g. poverty) which may be raised.

An increasingly popular means of establishing local views, aspirations and opinions is through 'focus groups' of selected members of the public, who are assumed to reflect the views of local socio-economic categories (e.g. Krueger, 1994). Whereas quantitative methodologies tend to force responses into predetermined categories, focus groups exemplify a qualitative approach with more open-ended responses. They attempt to reflect the intricate processes of reasoning and argument involved in the discussion of topics of public concern, and also provide an opportunity for exploring the wider social frameworks in which particular issues are discussed. Focus groups consist of six to ten participants, generally discussing a particular topic for around two hours. Whilst the participants are not intended to be 'representative' in a statistical sense, they are assumed to typify characteristic groupings. Focus groups are carefully mediated by an experienced moderator, who facilitates a group dynamic in an open and non-judgemental fashion, so that the group is able to explore the full range of meanings and understandings of a particular topic. The moderator introduces issues and generally 'feeds' the discussion, though is careful not to present personal opinions (Mcnaghten et al, 1995). This approach is being increasingly widely used to evaluate public opinions towards the environment. Applications have included:

- evaluating public attitudes towards and knowledge of key environmental issues;
- identifying features of the environment which particular groups enjoy, cherish or fear;
- calculating economic 'values' for the shadow prices which people would hypothetically 'pay' for unpriced environmental 'common goods';
- determining environmental indicators which reflect topics of importance to the affected public; and
- evaluating the public's attitude towards public environmental agencies, such as local government.

Networks and partnerships

The retreat from simple, hierarchical, top-down, command-and-control methods of environmental management has popularised the idea of 'networks' of partners collaborating to produce sustainable solutions. Although the alternative model of bottom-up, grass roots management is now generally accepted to be too simplistic, the idea of equal partners helping to generate solutions and assume complementary responsibilities for their implementation remains appealing. These range from high-level 'policy networks' in which powerful government departments and agencies interact to negotiate their agendas and domains of action, to local self-help arrangements geared to more specific and immediate needs. Apart from their accessibility and inclusiveness, networks are popular mechanisms because they help members to 'manage across boundaries' in inter-disciplinary situations. They also extend the possibility of organisations using each others' invisible assets (know-how, local knowledge, past experience) at little or no extra cost (Martin, 1995).

Environmental networks have been associated with various types of promotion, advisory support and fundraising. They are especially valuable as means of disseminating information to target groups, as information received from known, personal contacts is normally perceived to be more reliable than information communicated from government departments or senior echelons of traditional, hierarchical organisations. Thus, partners within networks tend to benefit from the various connections which are cultivated, and may feel less vulnerable than they would if operating in isolation.

Networks also have their drawbacks as methods for local environmental management, and it is important not to see them as a panacea. Some critics have suggested that they merely assist the 'privatisation of environmental activity', detaching important decisions from publicly accountable bodies. Also, whilst one of their virtues is an ability to include lay members and volunteers, these can be exploited as unpaid sources of labour. Network members generally include some fairly small organisations who lack substantial financial resources and can thus find it impossible to pre-finance projects, or raise revenue to 'match' grants from government or private sponsors. More generally, network members may lack time and incentives to participate effectively, especially after the initial enthusiasm has worn off, or after a specific project has been implemented. Finally, much of the literature paints a rosy picture of networks acting amicably and selflessly, whereas in reality members often have conflicting objectives and agendas, and their operation can become tortuous.

Whilst many organisations develop mutual ways of working in relation to a variety of project and sponsorship initiatives, it is generally agreed that formal partnership should go beyond these loose working arrangements. Forrester (1990) has thus defined partnership as:

> . . . a recongition of mutual needs and interests which entails a longer and closer relationship than the provision of funds for a worthwhile promotional activity . . . (it is) continuous rather than sporadic.

Equally, observers concur that partners need a firm basis on which to work together (the term 'para-legal' has been used in some circumstances), rather than a pious hope. Some have even advocated the use of contracts with specific targets for performance by each of the agents. However, the broad intention is to encourage a process of collaboration, through which parties who are only usually aware of their own aspect of a problem can explore their differences constructively, and search for solutions, with other groups.

Effective partnership typically displays certain features, which may be associated with features such as relatively formal cycles of meeting, methods of reporting and arrangements for convening. Wood and Gray (1991) propose that they can be characterised in terms of:

- stakeholders – the groups or organisations with an interest in the problem domain (on a 'majority' or 'minority' stakeholder basis), and whose interests may change during the collaborative process;
- autonomy – whilst stakeholders must agree to abide by shared rules, they still retain independent decision-making powers;
- interactive process – the working relationship is oriented to change, and will itself change over time;
- shared norms, rules and structures – stakeholders agree on their *modus operandi*, but these rules and norms are typically temporary and will evolve or realign themselves over time;
- action or decision – collaboration must work towards an objective or intention; and
- domain orientation – the partnership must relate its work to a common problem domain, and its orientation must concern 'the future'.

Many types of joint working for sustainable development do not conform to these principles, for example statutory panels or temporary (and relatively anarchic) alliances introduced for 'catalytic' purposes.

Conclusion

Local sustainability strategies are increasingly being based on a suite of well-established methods and techniques. There is still a high degree of experimentation with appropriate methodologies: some local authorities are producing strategies based on a comparatively 'top-down' approach of expert decision-support methods, whilst others are adopting a 'bottom-up' approach to envisioning and consensus-building. Both of these overlap to a considerable extent and a combination would be appropriate to the 'sandwich' strategies described in the previous chapter.

This variety is to be encouraged, especially at the formative stage of sustainability planning. In the future, it is probable that a more standardised approach will emerge, especially as the theoretical and information base of environmental indicators improves, and as imprecise notions, such as 'capacity', become more clearly defined. However, given the socially constructed nature of our perceptions of environmental crisis, and the volatile outcomes of civic science and participative planning, it is likely that our selection of methods and techniques will be subject to constant adjustment and refinement.

5

Local government and sustainability

Introduction

The pursuit of sustainability is clearly a multi-partite endeavour, and it is not something to be achieved by governments acting alone. Indeed, some market research suggests that people are increasingly less respectful of their governments, and tend to mistrust politicians' motives and honesty. Governments do, however, have a particular responsibility to support sustainability programmes, both as creators of legislation and public policy, and as funders of major environmental programmes.

Central government, despite its legislative and economic powers, is rarely in a position to instigate direct change at the local level. With regard to individuals and communities, therefore, it is likely that local government will be of greater significance. The local government system comprises a range of councils, some of which have split responsibilities (i.e. powers are divided between county and district/borough) and others which are all-purpose in their functions. Below this tier are parish, town and community councils which possess a very restricted range of functions, but which can lead a number of important initiatives relevant to local pride and citizenship. One role of local authorities, therefore, is to be responsive to citizens' needs at the local level; they are major providers of services in people's daily lives, and are democratic bodies, accountable to the local electorate. Analysis of Agenda 21 has suggested that a high proportion of its action proposals fall within the domain of local government.

The environmental role of local government

Since 1980, most political commentators have claimed that the autonomy of local authorities in Britain has been eroded. Principal reasons have been tightening financial constraints and the requirement for a large range of local government services to be 'contracted out'. This latter generally takes place on the basis of 'compulsory competitive tendering' (CCT), through which contracts are let mainly to private companies and trusts. In a sense, therefore, local government is in a weaker position to be generous in its support of environmental initiatives, and has lost direct control over some of its 'environmental' activities. Conversely, it could be argued that the new régime of local government, which is increasingly concerned with the quality of services delivered to the public, is highly compatible with the culture of total quality management. It is this culture which has influenced the design of environmental management systems and 'green audits' so that, in theory at least, there is a good deal of fertile ground within which local government environmental quality assurance can take root. Equally, local authorities are in a strong position to be standard-bearers of responsible environmental management, and the pursuit of sustainable development at the local level may become central to the *raison d'être* of local government.

Whilst many councils will find it difficult to support specific environmental projects, therefore, they may become exemplars of the ways in which an organisation can behave responsibly in relation to its use of resources and production of wastes. Indeed, this will be essential if they are to exercise civic leadership, and catalyse wider environmental citizenship. This requires 'greening' at a much more fundamental level than merely adding on a few newsworthy projects.

One of the principal advocates of the importance of local authorities in Britain has been the Local Government Management Board. Its recent publications (e.g. LGMB, 1992a,b; Stewart and Hams, 1991) have included abundant guidance on environmental management issues, and these set out various reasons why local authorities are key players in sustainable development. First, they are reckoned to be close to many of the major issues, such as land-use planning and solid-waste disposal. Second, they have a substantial potential for 'capacity building', namely, providing people with the knowledge, powers and resources to undertake sustainable development. Third, they are the source of local democracy, thus providing 'local choice' as well as a 'local voice', and a basis for citizen participation. Fourth, they are well placed to be sensitive to local identities and needs, and can therefore accommodate local differences, diversity and innovation. Councils also have an understanding of the scope for local initiatives, and the impacts that development may have on particular areas. Fifth, they can try to create the conditions in which local action can commence and flourish, and may themselves be a starting point for new initiatives. More widely, they can be a partner in initiatives which range across districts, regions, counties and even countries. Bosworth (1993) has also pointed to the responsibilities of local authorities as big resource users in their own right, and as the level of elected government closest to the citizen, with the clearest local democratic mandate.

At least as significant as the representative and leadership role of local government is its statutory responsibility to implement various areas of legislation which have direct or indirect environmental consequences. These include town and country planning, economic development, housing management, highways and transport, control of certain pollutants, solid waste management, and education. As previously noted, some of these are entirely or partially contracted out to trusts or private sector companies, whose environmental integrity may be 'chained' through CCT and quality assurance systems. There is a recognised need to ensure that the environmental consequences of service provision are reflected in contracts. Appropriate contractual clauses can also ensure that environmental responsibility is 'chained', namely, that service providers must also ensure that sub-contractors and suppliers have satisfactory environmental management systems, although there are delicate legal issues regarding local authorities' abilities to restrict open competition by imposing environmental requirements.

Broadly speaking, local authorities are responsible for two types of impact on the environment: direct effects and service effects. Direct effects are the environmental consequences of the production of goods and services which arise from their own day-to-day activities. These include the consumption of energy and water in council-owned buildings, and by their plant and equipment. They also include the use of transport by council staff, and the ways in which this contributes to fuel consumption, emissions and traffic congestion. They can refer to the consumption of resources and materials through purchasing procedures, and the solid wastes and polluting emissions and discharges which arise from council activities. All of these are relatively easily measurable. More difficult to determine, but probably even more significant, are the 'service effects' associated with the services which councils provide and the policies which they adopt. Some of these are relatively focused and quantifiable, such as road construction, air pollution regulation and recycling programmes; others are more diffuse, such as town planning policies and provision of consumer information on 'green products'.

The greening of local government

The rising importance of green issues in local authorities is truly remarkable. During the 1980s, few paid serious attention to environmental management and, if they considered it at all, responsibility for such matters was likely to be located in a section of a department well away from the main power base of the council. By the end of the 1980s, interest was starting to grow. A proposed method of State of Environment Reporting, developed by Friends of the Earth, was tested out in Kirklees Metropolitan Borough Council (KMBC, 1989). Raemaekers (1993), reporting on results of surveys of local authority 'green plans' in 1991 and 1992, was able to confirm a solid, if unspectacular, increase in the level of activity. In the latter survey, of the 87% of councils which responded, 74% were 'active' in the sense of having produced some sort of green plan, though only two per cent met the most rigorous criterion of a full suite of specific environmental documents.

There are many valid models for the ways in which green issues are managed within a council, and the ways in which these are brought to committee.

Nowadays, however, it is increasingly likely that environmental responsibilities, including the production of a regular and systematic audit of a council's direct and indirect activities, will be located in the Chief Executive's Department and that the environmental implications of policies will be a regular item on the agenda of all committees.

The reasons underlying the growth in significance of environmental issues at local government level has been analysed by Ward (1993), who interpreted it in terms of various models of the ways issues emerge as contenders for serious attention. These, he referred to as:

- the 'outside initiative' model, which assumes that individuals and groups external to the main organisation (in this case, local government) pressurise policy makers to adopt their concerns and solutions;
- the 'mobilisation' model, where political leaders initiate policies, and then attempt to secure popular support for them;
- the 'inside initiative' model, where policy entrepreneurs within the organisation try to promote their objectives, but do not wish to 'go public' on their ideas in case they do not win popular support, and
- the 'convergent voice' model, in which the same issue is independently articulated by different groups inside and outside the organisation at about the same time.

Ward's evidence broadly suggested that the greening of local authorities was a response mainly to environmental activists, who wanted to see environmental policies addressed in a holistic and corporate fashion, rather than in a merely incremental fashion. In this sense, the appropriate model was one of 'outside initiative', in which groups mobilised their resources to force environmental issues as a package up the local government agenda. However, during a period of waning powers for local authorities, councils were often glad to seize on this new agenda and the new role which it gave them. Ward considered that local authorities sought to re-define environmental issues for a variety of their own purposes, not least to strengthen their importance and popularity relative to central government. In this regard, there were resonances with the convergent voice model.

In practice, the situation proved to be rather more subtle than these generalised models might infer. Re-definition of environmental issues could have suggested that local government was merely repackaging its traditional duties and activities in terms of the fashionable jargon of sustainable development. Whilst there is more than a grain of truth in this, equally, there have been some genuinely new departures in integrated operations and corporate management. Also, some important new knowledge and skills networks have emerged, such as the Environment City programme and the International Council for Local Environmental Initiatives. One body of research on government has focused on the emergence of 'policy networks' between departments and agencies who try to negotiate informally to 'set' the agendas which they want, and thus generally behave as forces of policy inertia. However, in Ward's analysis it appeared that the new environmental networks were genuinely radicalising agents, aiming for increased participation and pressuring for implementation. Conversely, the production of 'green plans' in local government may serve to institutionalise a particular set of policies and establish a bureaucratic routine for environmental management.

Overall, environmental interests in local government appear to be characterised by a general level of commitment, overlaid by occasional party political differences on the funding of specific initiatives, and small networks of more active policy entrepreneurs working across traditional party lines. Perhaps, though, the more significant consequence of these observations is that local authority sustainability strategies take place within specific institutional, organisational and political contexts, and so the broadly rationalistic approach adopted by this book towards methods and procedures should be tempered by the more qualitative influences which will frame initiatives in actual situations.

The generally favourable conclusions of many analysts (some of whom are, in practice, advocates for local government) must be tempered with the difficulties experienced in implementing substantive local programmes. Moore's analysis of the problems faced by the Vancouver Task Force on Atmospheric Change identified a number of barriers to effective action (Moore, 1994). Her survey of officers, elected members and citizens drew attention to three principal sources of difficulty and delay. First, a cluster of 'perceptual/behavioural' barriers affected the individuals within the responsible organisation. These included inadequate comprehension of environmental policies, perceived lack of effective powers, competing issues for attention and finance, and lack of 'buy-in' (i.e. feeling of personal commitment to and ownership of the programme) by civic staff. A second set of 'institutional/structural barriers' affected the organisation itself, such as fear of losing constituent support and inappropriate government structures (essentially, bureaucratic difficulties of working cross-departmentally in a 'horizontal' fashion). Finally, 'economic/financial' barriers were found to affect the organisation's ability to function within a wider societal context. The main aspects were inadequate funds and other resources, and a failure to guarantee results. Moore's findings resonate with Trudgill's more familiar 'barriers' to environmental problem-solving (Trudgill, 1990), namely:

- agreement (lack of consensus over the nature of the problem)
- knowledge (inadequate or conflicting information base)
- technology (lack of effective technical solutions)
- economic (available solutions are too costly or relatively unattractive)
- social (lack of social acceptability, perhaps because of perceived inequality or personal sacrifice)
- political (lack of political will).

These considerations introduce a strong vein of circumspection to our review of local government's environmental role, especially when confronted with complex and disputed problems, and must temper some of the currently fashionable rhetoric.

Table 5.1 Summary of issues covered in the 'model' *local environmental charter*

Topic Issue	*what you can expect* *(examples only)*	*good practice* *(examples only)*	*how you can help* *(examples only)*
access to environmental information	have a right to see various papers on council decisions + public registers	giving opportunities to see information outside normal office hours	
air pollution control	right to see and check public register	giving daily bulletins on air quality	telling council if you think there is an air pollution problem and providing a detailed description
beaches	council must keep popular beaches free from litter (including dog mess) between May and September	displaying standards and main results of bathing water quality tests in prominent positions on the beach	reporting cases of oil pollution
common land	maintain registers which show details of land, rights and ownership		
dog nuisance	council must collect stray dogs	providing dog waste bins	telling council if you know of an area where dog mess is a problem
drainage and flood defence	if an area is liable to flooding, council may have powers to build flood defences	responding to complaints about leaking private sewers within four hours	
drinking water	can ask council to test the quality of drinking water	provide information on grants for replacing lead pipes	
licensing	can comment on any proposals for licences	keeping record of any serious problems	
litter	can declare 'litter control areas' and issue 'street litter control notices'	appointing litter wardens in areas where there are special problems	telling council or police if witness fly-tipping
noise, smoke, smells and other nuisances	council can take action if problem constitutes a 'statutory nuisance'	using noise mediation services to solve problems between neighbours	trying to resolve problems with neighbours 'reasonably'
obstructions in the road	temporary obstructions must be authorised	rapid response to complaints about obstructions	reporting obstructions which are causing damage to the road or path

pest control	council must keep pests under control in their own property	responding to rodent problems within three days and to insect problems within five days	
planning	have right to comment on the 'development plan'	providing wide opportunities to see planning proposals, e.g. through public libraries	
recycling	council must prepare and regularly review a 'recycling plan'	setting a target of household waste which it is aimed to recycle	buying recycled products
rights of way	council is responsible for maintaining rights of way and for signposting them	responding to complaints about rights of way within 14 days	
roads and pavements	have duty to maintain their pavements and roads properly	protecting roadside wildlife, e.g. by not mowing verges until wildflowers have seeded	reporting any dangerous obstructions in the road
rubbish collection	have right to have everyday household rubbish uplifted	removing any fly-tipping within 24 hours of being told about it	
traffic management and parking	can regulate all kinds of traffic on public roads	using vehicles which run on unleaded petrol and encouraging their staff to use public transport	reporting any smoky vehicles
tree preservation	have the right to ask the council to protect trees of special value	providing advice about trees and tree care when asked	planting new trees
waste regulation	licensing land fill sites	responding to complaints about licenced contractors within two weeks	
wildlife areas	right to make points about nature conservation on individual development proposals	adopting a nature conservation strategy	reporting damage/ disturbance to wild animals or plants

Source: based on CALGEFWG, 1994

Environmental reporting and planning

One of the principal roles which local government can play in environmental management is that of providing information to the local community. Environmental information is, however, generally fragmentary at the local level, and is rarely easily interpreted by the general public. Local datasets thus have to be transformed into genuinely informative reports and displays. An equally important role is for local authorities to communicate their environmental responsibilities and performance targets to local taxpayers. Consequently, a variety of types of document has emerged, each serving particular functions. Broadly, they are of four types:

- *environmental charters* – setting out the performance standards which a local authority aims to achieve on 'green' issues, as well as the ways in which they expect to support, encourage, review and promote action by individuals and groups;
- *'green plans'* – 'informal' plans setting out a range of policies and action statements on a corporate basis, and across the local authority;
- *state of environment reports* – 'external' audits of ambient conditions within the geographical area administered by the local authority; and
- *local authority audits* – 'internal' audits of the direct and indirect environmental effects of the authority itself.

These documents have been of very variable quality, some comprising detailed policies and commitments, others conveying only broad aims and generalised budgets and lacking deadlines. The Central and Local Government Environment Forum Working Group (1994) has attempted to set standards for the coverage and detail of Environment Charters (Table 5.1). These standards are rather narrowly defined, and are largely devoted to the provision of available information, the discharge of duties in relation to statutory or advisory standards, and the ways in which citizens can inform and assist their councils. Without being especially radical, the CLGEFWG document does nevertheless indicate ways in which local good practice can exceed a basic minimum.

The term 'green plan' has been used very broadly to describe a non-statutory report which covers a council's environmental policies and action programmes. It should not be confused with the 'development plan' which is a statutory document covering a stipulated range of land use issues. Documents which are entitled 'green plan' or 'environmental action plan' typically elaborate on the commitments contained in a council's environmental charter, and are normally corporate (i.e. council-wide) statements about the many actions and responsibilities spanning local authority departments. Characteristically, they contain a large number of short-term but high profile project proposals capable of providing a rapid return political return; however, they may include more challenging targets which can only be met over the longer term.

One example of this is provided by the Wakefield green plan (Seddon, 1994), which focuses mainly on ecology. In an area characterised both by high

quality amenities and environmental degradation, the plan's five key objectives
are to:

- identify and secure a network of green spaces;
- identify areas where the Council can use its influence to conserve, enhance
  or develop landscapes for wildlife and for public use and enjoyment;
- encourage landowners and managers in the public and private sectors to
  adopt ecological approaches to environmental planning and management;
- ensure that residents of the district are within a half a kilometre of an area of
  ecological interest; and to
- promote enjoyment in the natural environment and consideration for its
  wildlife resources.

The main value of the plan has evidently been to move away from a 'protected
areas' approach, which merely helped the Council react to development pressures, towards one of strategic promotion of a network of wildlife corridors and
areas, enabling the Council to anticipate the most effective ways in which the
area could benefit from development.

Internal audits

The most systematic attempt to integrate environmental decision-making into
local government has been through 'internal audit' procedures, in particular
the Environmental Management and Audit Scheme for Local Government
(EMAS) (HMSO, 1993). This is an evaluative framework developed by LGMB
and others, on the basis of the European Union's EMAS scheme for industry. It
is a voluntary, rather than a compulsory, scheme which, after being piloted in
seven UK local authorities, many others have chosen to enter. Its aim is to
describe and quantify an authority's direct and service effects on the environment, on a corporate basis, and to help it continuously to improve on its
current performance. Proponents of the scheme envisage that participating
authorities will realise a range of efficiency gains, and not simply environmental benefits. The basic approach is threefold:

- the establishment and improvement in local authorities of environmental
  policies, programmes and management plans;
- the systematic, objective and periodic evaluation of these policies, plans and
  management systems; and
- the provision of information on environmental performance to the public.

It is important to recognise that the scheme does not prescribe particular
environmental policies or performance standards, but establishes a management framework to instil good and continuously improving practice (i.e. it
aims to provide quality assurance rather than quality control).

The starting point of the scheme is a set of worksheets (Figure 5.1), accompanied by extensive guidance, which provide the factual baseline and conceptual framework for the exercise. *Service effects* are broken down into

WORKSHEET: SCOPING REVIEW OF DIRECT EFFECTS

Operational unit
Worksheet filled in by (name)   *on*   *(date)*
Energy use in buildings, plant and equipment
1. Does the unit use – and control the use of – significant amounts of energy in its buildings, plant and electricity using equipment?

   (a) please list the buildings
   (b) please list the plant
   (c) please list the equipment and what it is used for

Water consumption

2. Do any of the activities of the unit use significant amounts of water?

   Please list these activities

Transport use

3. Do any activities of the unit make use of transport vehicles or petrol or diesel-using plant and equipment to a significant degree?

   Please list these activities, and the type of transport, plant or equipment they use.

Purchasing

4. Does the unit make any purchases whose types or quantity could result in significant environmental effects?

   Please list these purchases (in groups of related items), and the activities for which they are purchased

Wastes and pollution

5. Do any activities of the unit produce significant amounts of solid waste of significant quantities of emissions to air or discharges to water courses, drains or sewers?

   Please list these activities, and the type of wastes, emissions and discharges produced

Other direct effects

6. Does the unit have any other direct effects on the environment, for example, through noise or new buildings?

   Please list them

Figure 5.1   Example of a Worksheet in the Local Government EMAS manual. This particular one refers to the 'Scoping Review of Direct Effects' (i.e. as opposed to 'service effects'), which serves as a starting point for worksheets on more detailed aspects.

Source: HMSO, 1993, Crown Copyright.

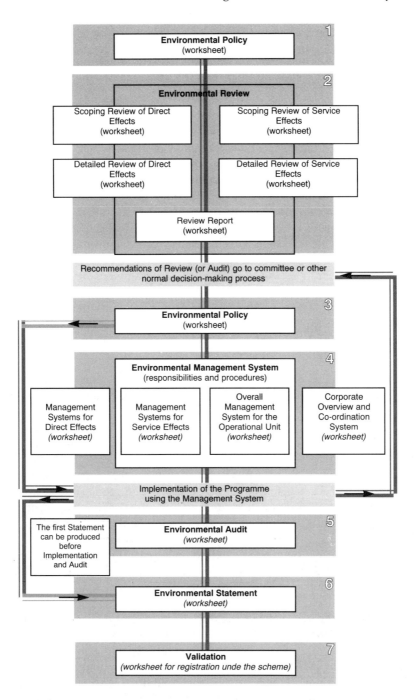

Figure 5.2 Stages in the local government EMAS procedure
Source: HMSO, 1993, Crown Copyright.

constituent activities, and a 'scoping review' is conducted on each of them to select those with significant environmental effects. *Direct effects* often derive from centralised procedures, and are thus analysed for an operational unit or local authority as a whole, rather than for disaggregated activities. The main stages (Figure 5.2) of the procedure comprise:

- the organisation's environmental *policy*, which must go beyond compliance with minimum legislative standards;
- a *review* of the environmental impacts of the activities being considered (and their regulatory and policy context);
- a *programme* of activities to achieve defined *objectives* which translate the policy's aims into specific quantified goals for improvement (these typically make use of 'environmental indicators' or target levels of performance);
- a *management* system which defines responsibilities, procedures and tools for implementing the programme;
- periodic *audits* to assess whether the programme is being followed and whether the management system is adequate;
- a *statement* of environmental performance, which must be published; and
- impartial external *verification* of the process, leading to a formal *validation* of the public statement.

Throughout the whole process, involvement of the general public – through information provision, consultation and handling of complaints – is a pervasive feature.

This is clearly an ambitious corporate task, and well beyond the scope of one person, yet it must emerge as a cohesive exercise. One approach has been to appoint one person within the authority as a 'green officer'; this has the disadvantage that they are likely to be detached from the main decision-making power base, yet the benefit that they may have a high profile, and be able to act as a catalyst. More commonly, a co-ordinator of the eco-management process is appointed whose responsibility is to cascade actual tasks to small groups of specialist officers, thereby both expanding the base of expertise and broadening the sense of 'ownership' and commitment amongst staff. Accredited verifiers, usually from consultancies or verification bodies, are then normally brought in to confirm independently that the management system is working as claimed.

The main hallmark of EMAS is that it is a *corporate* activity, so that the environment becomes part of the mainstream business of the authority rather than an ornamental extra. The corporate review and coordinating system associated with EMAS requires that a local authority makes various provisions. These include a committee or sub-committee with responsibilities for overall formulation and monitoring of the environmental strategy and action programme. It also requires that a named chief officer has a clear responsibility for environmental management, and that there are well-defined procedures for inter-departmental coordination. Also, there must be procedures for the environmental review of budgets and a corporate environmental information system.

The overall approach is similar to the environmental management systems (EMSs) being introduced in industry (see Chapter 6) but differs in that it is much more exposed to public scrutiny. Thus, Davis (1994) has noted that EMAS is distinctive in the extent to which it requires authorities to produce regular public statements which report progress against agreed targets. It is anticipated that the regular exposure to public scrutiny will provide the political and institutional stimulus to achieve the continuous environmental improvements which are the essential intention of the scheme. Davis also noted a number of specific benefits arising from EMAS, namely:

- continuous demonstrable improvement in an organisation's activities;
- flexibility, in terms of conceiving solutions which are appropriate to specific causes;
- ability to assist in addressing environmental issues in a systematic manner, and in setting priorities;
- an ability to improve the quality of any subsequent environmental strategies;
- a high public profile, which helps to maintain the popular credibility of a local authority;
- a potential to save costs; and
- an ability to take an anticipatory view of the demands of European legislation.

An example of a local authority environmental audit is that conducted by Staffordshire County Council (1992), which furnishes a threefold examination of the county's policies in terms of their environmental effects. The first part of the audit contains an *assessment of services*, which looks at the environmental impact of council policies and practices which directly affect the public and the 'outside world'. The second part contains an *assessment of internal practices*, looking at the impact of the practices which the Council's departments have on the external environment. Finally, there is an evaluation of the corporate management system of the council, in terms of its delivery of environmental management services and maintenance of environmental standards. Each section contains a set of 'white pages', which describe the characteristics of topics relevant to the environment, 'blue pages' relating topics to departmental responsibilities, 'yellow pages' which relate the Council's current policies and practices to prevailing 'best practice', and 'green pages' which suggest ways in which the Council's performance could become more environmentally friendly.

The audit considers a spectrum of topics, some under the Council's control and other where its role is very indirect, but still related to issues of public interest (Table 5.2). For example, in relation to transport, the audit reviews the Council's location with respect to the national communications network, and

Table 5.2   Subject areas in Staffordshire's environmental audit

- development;
- transport;
- heritage (landscape and buldings);
- nature conservation;
- agriculture;
- recreation;
- waste and recycling;
- public participation;
- education; and
- information, liaison and community involvement.

Source: Staffordshire County Council, 1992

refers to national concerns about traffic congestion, the impact of highway construction, and associated demands for aggregates. It then reviews the legislative powers (e.g. Road Traffic Act 1988) and the responsibilities and roles of departments and committees (e.g. Highways Committee, Transportation Sub-Committee, and Joint Transportation Planning Team), followed by an analysis of the Council's current policies and practices (e.g. local studies of public transport need, highway schemes). County Council performance is then compared with best practice elsewhere, showing, for instance, that they compare favourably on their area-based assessments of transport needs, but have opportunities to be more effective in relation to multi-use transport schemes. Finally, consideration is given to the increased use of environmentally friendly approaches, such as traffic demand management. In areas where the Council's impact is only marginal, such as agriculture, the audit focuses more on ways of influencing the major stakeholders, through information, advice and negotiation.

Local Agenda 21

The text of Agenda 21 enjoined governments to produce 'national sustainable development strategies' (NSDS), and this has been widely undertaken. Alongside this, as previously discussed, local government organisations have produced 'local sustainable development strategies', generally known as Local Agenda 21s. The body of experience in preparing strategies either at the national or local level is not yet very great and there has been a considerable amount of diversity and experimentation. The range of international projects was extensively discussed at a major conference in Manchester in 1994 (Whittaker, 1995). Perhaps the most striking lesson to emerge from this discussion was the differing concerns of less developed and more developed nations. The former generally emphasised fundamental needs, such as instilling more democratic local processes and improving basic planning and infrastructure services, with a concern for 'brown' issues (dereliction, waste disposal, sewerage, etc.), while the latter tended to display a greater concern for reducing excessive resource consumption and for 'green' issues (amenity, landscape conservation, etc.).

Table 5.3 Factors influencing success in partnerships for sustainability and Local Agenda 21 programmes

Positive features in environmental partnerships:
- need for a specific focus
- multi-partite comprehension of the nature of the problem
- solutions which are appropriate to the context
- approaches which are innovative and flexible (i.e. a willingness to adapt and to avoid mindset)
- inclusion of diplomacy and conflict resolution skills
- inclusion of 'animateur'
- proposals to build capacity (individual and organisational)
- deliberate diversity
- adequate financial resources
- commitment to communication
- ownership of the partnership by all parties
- wide participation
- trust, transparency and accountability
- north-south dimension
- leadership and clarity
- evidence of added value and specific projects
- experts 'on tap' rather than 'on top'

Negative features in environmental partnerships:
- hidden agendas
- inequality, competitiveness and intolerance
- sectoralism
- excessive dependence on external aid/expertise
- inadequate administrative support
- problem avoidance (acceptance of a 'false consensus')
- mutual distrust
- different 'languages' of different sectors
- poorly developed methodology
- sharp changes to existing structures
- excessive complexity
- over-reliance on experimental approaches

Source: based mainly on information in DoE, 1994b and Freeman et al., 1996

Despite the diversity, we may start to make some helpful observations about sustainability strategies. Perhaps not surprisingly, it appears that the best ones have been based on widespread participation, and have built on good existing plans and processes, rather than trying to start from a zero base. Moreover, they tend to give clear attention to developmental or economic, as well as environmental, priorities. Conversely, failed strategies tend to be prepared by small élite task forces without consultation, neglecting existing initiatives, and having a limited scope (Carew-Reid et al, 1994). An evaluation of partnership approaches to environmental management (DoE, 1994b) identified a similar range of positive and negative factors. The better features reflect an open, innovative and multilateral approach. Counter-indicators include intolerance, mistrust, sectoralism, and a tendency to avoid rather than resolve contentious issues.

Two observations are of particular importance. First, strategies need to look at social and personal needs as well as environmental ones; especially in poorer countries, this means integrating developmental needs with ecological priorities. Second, strategies which deal with policy processes at the national level need to be linked to planning and action at the local level. Fulfilment of this latter requirement may depend on various factors such as:

- a location for its steering committee and secretariat where they can have the greatest influence on key decision-makers;
- high level support for corporate implementation of the strategy;
- the commitment of key participants; and
- a conducive political and social climate.

These are daunting prerequisites, and they do serve strongly to dispel the idea that LA21 exercises are merely about producing documents. Rather, they are about trying to achieve the right chemistry for acceptance and success.

The need to prepare LA21s has been endorsed with surprising enthusiasm, as reflected by the way in which many European towns and cities have affirmed the 'Aalborg Charter'. This reflects the commitment by a number of municipalities, in the wake of Chapter 28 of Agenda 21, to join the European Sustainable Cities and Towns Campaign. The Charter makes specific reference to the need for achieving local consensus, and making progress towards the objectives of the Fifth European Action Programme. At the heart of the Charter are eight stages in the production of a local action plan, namely:

1. recognition of the existing planning and financial frameworks as well as other plans and programmes;
2. the systematic identification by means of extensive public consultation, of problems and their causes;
3. the prioritisation of tasks to address identified problems;
4. the creation of a vision for a sustainable community through a participatory process involving all sectors of the community;
5. the consideration and assessment of alternative strategic options;
6. the establishment of a long-term local action plan towards sustainability which includes measurable targets;
7. the programming of the implementation of the plan including the preparation of a timetable and statement of allocation of responsibilities among the partners; and
8. the establishment of systems and procedures for monitoring and reporting on the implementation of the plan.

These principles are underpinned by a more general recognition of the need to review governmental processes and capacity-building frameworks.

A further high-level response has been that of the International Council for Local Environmental Initiatives (ICLEI), particularly through its Model Communities Programme and Strategic Services Planning Framework. The ICLEI recommends that LA21s in its selected 'model communities' (an international network of about a dozen municipalities) should instil a blend of effective technical systems and democratic processes. The Planning Framework is intended to deliver services in a way that increases resource conservation and maximises their long-term social and economic viability. A strategic action plan is included which sets multi-sectoral short- and long-term targets. The Framework's objectives are thus to:

- engage service users and civic partners in the design of municipal service strategies;
- identify and analyse systemic service problems and to fully incorporate long-term environmental, social and economic objectives, including measurable targets, into problem-solving plans;
- implement service action plans based on partnerships between the public sector, civic sector and business sector as well as individual service users themselves; and
- engage service users and civic partners in the ongoing evaluation of services and their sustainability.

Accompanying these are democratic and public communication elements, intended to ensure a self-sustaining community consultation process in which all sectors are fully informed and involved. Important elements include a sustainable development auditing method which allows a wide range of participants to assist local authorities in auditing current conditions, and a suite of sustainability indicators which are clear, meaningful and achievable. For inclusion in the programme, each Model Community must:

- establish a multi-stakeholder planning process, provide a formal mandate to a local Stakeholder Group and support the process for a minimum three-year period;
- constitute and provide an official mandate to a Project Team which will serve as the staff for the Stakeholder Group and its planning effort. The Project Team must include municipal staff and non-municipal professionals from environment, development, social services and/or health professions;
- create an interdepartmental committee of municipal department representatives to liaise with the planning process and assist in the development of the action plan; and
- formally adopt a sustainable action plan ('Local Agenda 21') within 30 months of project commencement.

As this is essentially a research programme, ICLEI is starting to accumulate substantial feedback and survey results on the effectiveness of municipal sustainability measures.

Early experience in Britain

Although there is a considerable degree of independence and experimentation in contemporary British experience, there is also a core of consensus. This has been aided by the Local Agenda 21 Steering Group, set up by the various 'local

Table 5.4   Key areas of action in the Local Agenda 21 process

Three areas for action within the local authority

Managing and improving the local authority's own environmental performance
- corporate commitment
- staff training and awareness raising
- environmental management systems and budgeting
- cross-sectoral policy integration

Integrating sustainable development aims into the local authority's policies and activities
- green housekeeping
- land use planning, transport, housing and economic development
- tendering and purchasing
- tourism and visitor strategies
- health, welfare, equal opportunities and poverty strategies
- explicitly 'environmental' services

Awareness-raising and education
- support for environmental education and voluntary groups
- visits, talks and awareness-raising events
- publication of local information and press releases
- initiatives to encourage behaviour change and practical action

Three areas for action within the wider community

Consulting and involving the general public
- public consultation processes
- forums, focus groups and feedback mechanisms
- 'planning for real' and parish maps

Partnerships
- meetings, workshops, conferences, roundtables, working/advisory groups
- Environment City model
- partnership initiatives
- developing-world partnerships and support

Measuring, monitoring and reporting on progress towards sustainability
- environmental monitoring and state of environment reporting
- sustainability indicators and targets
- environmental impact assessment and strategic environmental assessment

Source: based on UK Local Agenda 21 Steering Group, 1994, cited in Whittaker, 1995

authority associations' (i.e. organisations representing the collective interests of counties, districts, etc.), and including nominees from industry, trade unions, the voluntary sector, women's groups and higher education, as well as local government. The Steering Group has organised conferences and a round-table, and has prepared various guidance notes. In particular, it has publicised six critical areas in the Local Agenda 21 process (Table 5.4).

Despite (or perhaps because of) the potentially challenging nature of a sustainability strategy, Pritchard (1995) reported that just over seventy per cent of UK local authorities responding to a survey were committed to preparing a LA21, and this was a considerable increase on the previous year. Replies to this survey revealed widespread success in including sustainable development into the council's own corporate activities ('green housekeeping'). Perhaps not surprisingly, land-use planning was most often named as the area most susceptible to the incorporation of 'sustainability' policies. However, actions also ranged widely across housing, tourism, economic development, social services and education. This rapid uptake was attributed to the significant pre-existing level of environmental activity, so that LA21s were able to build on existing capacities rather than have to adopt a completely new way of thinking. A strong feature of the process was the level of co-operation between local government, Whitehall and environmental groups.

Not all the feedback from this survey was positive, however. Many councils were still not participating, citing problems such as lack of finance or staff time, lack of corporate commitment and the uncertainties of impending local government reorganisation. There was some concern that these views were simply masking deeper conflicts within local authorities. Also, several councils continued to perceive a real tension between economic development and sustainability and, whilst content to display a superficially green image, were nevertheless pressing ahead with environmentally-unfriendly industrial and road-building schemes. Overall, it appeared that there was considerable progress on short-term, high impact exercises, such as recycling schemes and the introduction of EMAS, though the survey alluded to deeper issues associated with long-term commitments to radical change and, perhaps, a waning public confidence in local government.

Some councils, nonetheless, have stolen an early lead in developing ambitious LA21s, and these were generally where a longer-standing commitment to integrated environmental management was already imbued. Four examples are reviewed briefly here, but the early stage of the process at the time of writing should be borne in mind: it is likely that all LA21s will mature rapidly and will be subject to extensive alteration. These accounts do little more than summarise the state of play at one point on the learning curve.

In the former Humberside County Council area, for instance, the environmental programme effectively commenced with the commissioning of a county audit in 1991, followed by the production of an Environmental Action Programme (EAP), which set out a democratically approved statement of future directions for County Council actions (Richards and Biddick, 1994). Subsequent annual action programmes confirmed that significant implementation targets had been met. The EAP was audited independently by a group of voluntary organisations known as the Humberside Environmental Network. The LA21 programme then diversified from this base to comprise five separate Action Programmes focusing on the Humber Estuary, wildlife, energy, education and waste. Various types of information were being provided to the programme organisers by several Standing Conferences, which both resulted in sharp exchanges of views, but also helped build constructive and positive partnerships. In all, the programmes coordinated around 900 initiatives by over 200 organisations, and a wealth of topic reports were submitted to the

County's environment sub-committee. Some of these initiatives have been continued by the successor authorities.

One of the most comprehensive of the county strategies has been that of Lancashire. This started most visibly with its green audit (LCC, 1991), notable at the time for the local political will which enabled a widely-envied sum of money to be allocated to the initiative. Arising out of the audit, Better Environmental Practices (BEPS) working groups were set up in each Council Department. One of the principal ventures arising from the programme has been the Lancashire Environmental Action Programme (LEAP), launched in May 1993. This is 'owned' by a forum (Lancashire Environmental Forum) comprising 90 member organisations. Themes emerging from LEAP (see Table 5.5) are fed back into council policies. Environmental good practice outside the council is promoted through a network of nine Centres of Environmental Excellence, which build on parts of the LEAP strategy and provide links with other community groups. Most of the centres are run by organisations other than the County Council, and are intended to develop complementary specialisms. Although lacking statutory powers, the Centres nevertheless enjoy considerable influence through information exchange, partnership, networking, education and awareness raising. Policy formulation within the Council has been aided by four Sector Working Groups covering:

- air, energy, transport, noise;
- water, waste, land, agriculture;
- wildlife, landscape, townscape, open space; and
- education and public awareness.

Future work is likely to involve a broader cross-section of the public, and to widen the sustainability debate to include more general 'quality of life' questions, such as fear of crime or worries about employment. In 1995, Lancashire's commitment to the environment was recognised in its selection as the only British authority to be included in ICLEI's Model Communities Program.

At the level of the individual municipality, perhaps the most significant 'green network' has been that established under the *Environment City*

Table 5.5   Themes within the Lancashire Environmental Action Plan

- Partnership for Action
- Tackling Global Warming
- Reviving Towns
- Cleaner Air
- Cleaner Water
- Protecting Land
- Reducing Waste
- Conserving Wildlife and the Countryside
- Raising Awareness

programme. This comprises a small network of cities committing themselves, with the aid of business sponsorship, to 'green' management, and it has significant influence both on project-related activity and on policy formation. One of the network members is Leicester, where a synergy between Environ (the charity responsible for coordination of the Environment City strategy) and the City Council has transpired. Within the Council, an Environmental Policies Working Party and Sub-Committee were established with the purpose of improving 'quality of life', promoting positive attitudes towards local and global environmental problems, and delivering practical projects which meet needs and reflect sustainable development principles.

Once again, the city's Local Agenda 21 programme (called, in this case, Blueprint for Leicester) has built on previous experience. As part of its Environment City status, Leicester had set itself the task of becoming 'a model of environmental excellence', committed to translating the concept of sustainable development into city-wide action. Public participation has been stimulated by a combination of questionnaire survey, visioning and community involvement. Many specific projects have been sponsored and the principles developed, and an idea of the range is illustrated in Table 5.6. Each of these areas is supported by a specialist working group (SWG), and guided by representatives from the local communities and businesses. The campaign has strongly emphasised the principle of 'partnership' between public, private and political sectors, which is seen as essential to create radical policies and programmes necessary to achieve fundamental change.

Table 5.6 Main topic areas and sample projects in Leicester Environment City programme

Transport	– continued commitment to expenditure on cycling over ten-year period; progressive 'Transport Choice' policy to reduce travel demand; and improve public transport.
Waste	– kerbside recycling going city-wide after extensive piloting; and recycling plan sets targets of 35–40% domestic waste;
Pollution	– CFCs removed from fridges, etc.; and air pollution monitoring system;
Energy	– City Council aims to halve its energy demand by 2025; and first British city to adopt 'Save Energy at Home' scheme;
Social	– Faith in Nature project; and 'Natural Curriculum' guide to link in with National Curriculum;
Ecology	– reclamation schemes; and comprehensive surveys/monitoring of open space and private gardens;
Buildings	– environment-friendly 'EcoHouse' run by Environment City charity helped publish a 'Building for the Environment' green development guide;
Economy	– local environmental business consultancy; and 'Environ Advisory Services' carried out many green business audits.

Source: Leicester City Council, 1995

Blueprint for Leicester was launched with a 'vision pack' seeking to develop people's vision about the city as a whole, and about their own neighbourhood. In developing the Blueprint's proposals, stress has been laid on achieving various types of balance. These include:

- balance between quality of life issues and environmental sustainability;
- balance between identifying public needs and hopes, and using specialists to meet these in sustainable ways; and
- balance between quantity and quality of participation.

As previously noted, the elusive notion of 'quality of life' has been central to the City's interpretation of environmental management, and it is this, rather than 'green issues' which dominates the Blueprint. Quality of life has been interpreted to reflect a sense of community which:

- supports individuals;
- supports a diverse and vibrant local economy;
- meets the needs of food and shelter; and
- gives access to fulfilling work that is of benefit to the community.

Thus, the LA21 programme:

- asks a range of important 'communities of interest' to take part; and
- targets 'priority groups' that have been under-represented in the past.

For this to work, it has been necessary to implement some quite radical changes in the Council's *modus operandi*, which have involved setting up new working arrangements with key organisations, ensuring commitment from senior officers within the local authority, and influencing the ways in which work plans and budgets are approved by officers and elected members.

A pioneering example of very locally-based action is the neighbourhood-level work undertaken in Reading. Here, the Borough Council entered a joint project, known as Going Local for a Better Environment (GLOBE), with the World Wide Fund for Nature. This was aimed at developing mechanisms to enable and encourage the participation of local 'community' groups in discussions regarding their current environmental concerns and the future environmental quality of the locality. These discussions covered the ways in which communities could contribute, and how their roles may be modified or altered to achieve realistic targets. Dialogues were intended to lead to the production of comprehensive plans for the future that took into account the needs, concerns and potentials of all contributing groups. They also addressed the information, insight and skills that groups would need to enable them to participate in, and benefit from, the achievement of new environmental standards.

Early action entailed identifying distinct neighbourhoods, and key organisations within these which would assist with local networking and provide support to neighbourhood 'facilitators' (who were specially recruited and trained local members of the community). Facilitators could then make contact with key people, groups and organisations in order to stimulate debate relating to local issues and needs. This process began to lead to the production of 'neighbourhood A21s'. Rather than being delineated on the basis of arbitrary or traditional administrative boundaries, neighbourhoods were defined according to 'maps of environmental activity', including, for example, the catchments of tenants' groups, recycling sites and 'Friends of . . .' groups.

The community development strategy at the core of the exercise entailed several steps, including:

- launch events (e.g. a public hearing, led by a trained facilitator);
- community liaison (a point of contact with a named officer in the Council);
- identifying local issues;
- broadening the issues (placing isolated community issues in the wider neighbourhood picture); and
- consensus and capacity building (associated with the development of community plans and neighbourhood A21s).

In the pioneer neighbourhood, a roundtable approach was adopted to 'develop a neighbourhood environmental vision', and this set priorities according to a 'SMART' procedure (specific targets, which are *m*easurable, *a*chievable, realistically attainable within likely resources, and *t*ime-limited), coupled with a SWOT analysis. In the Newtown neighbourhood, for instance, a review paper:

- listed the main environmental issues;
- categorised the main concerns;
- clarified action arising from priority categories;
- conducted a SWOT analysis;
- analysed responses to a community questionnaire; and
- set out draft action plans on topics such as open space, recycling and traffic.

Whilst the Council recognised that it was impossible for a community to be completely sustainable, it aimed to reduce communities' unsustainable behaviour and measure this reduction. Thus, a set of community-wide and town-wide environmental indicators was established. As the project matured, the process became integrated with the Council's well-established community development programme, and was able to adopt relatively familiar methods of neighbourhood development.

Conclusion

Contrary to the assumptions of some writers, the role of local government in environmental management is not synonymous with the concept of 'local sustainability'. However, its role is extremely important: because local authorities are local, they have authority, they can promote civic pride and leadership and they are guarantors of local democracy. Strategies for local sustainability are now becoming dominated by Local Agenda 21 exercises and these have invariably involved some input from local government, which does not always exercise a controlling interest. Approaches to LA21 have been largely experimental in the wake of Rio, but there are some elements of concept and organisation which are emerging as general 'good practice'.

One aspect of concern is whether the LA21 process can be fully democratic. Whilst a 'participatory democracy' model of widespread public involvement based on community action is becoming fashionable, there is a real risk that this process may be dominated by green activists and champions, and so fail to gain a genuinely broad base of support. Equally, the local government model of representative democracy, with its emphasis on members returned on the basis of infrequent and relatively poorly supported elections, may fail to inspire local citizens or harness their energies. If local sustainability strategies are to become genuinely sustainable, they must somehow combine the legitimation of local government with the vigour of community action.

The responses of local government to sustainable development are varied though some are ambitious. There is quite clearly a concern to move beyond rhetoric and superficial greening yet, at the same time, some ambivalence about the priority which green issues should be given. Some of the issues are probably irreconcilable, but others will benefit from a discursive consensus both within and outside the local authority itself.

6

Land-use planning and sustainability

Introduction

Town planning has a distinctive and central contribution to make to the pursuit of strategic and local sustainable development. Many of the conditions for local sustainability, especially the 'service effects' component of the EMAS audit, will need to be set through the land-use planning system. Agenda 21 and its subsequent guidance on preparation of national sustainable development strategies pre-suppose the availability of an effective planning framework. In countries where one does not exist, some kind of land use regulatory system needs to be installed as a means of improving urban conditions and building a capacity to respond to environmental pressures. Even in relation to rural resources, Agenda 21 expects there to be an integrated approach to planning and management, and advocates the use of 'landscape ecological planning'. In practice, though, integrated non-sectoral measures for rural resource planning are generally poorly developed.

Agenda 21 sees land-use planning as a prerequisite for 'promoting sustainable human settlement development'. In advanced industrial and post-industrial countries, urban management is such a familiar and well-developed activity that it is instructive to remind ourselves of the basic reasons for its existence. Agenda 21 defines the need for effective urban administration and capacity building as comprising:

- providing adequate shelter for all;
- improving human settlement management;
- promoting sustainable land-use planning and management;
- promoting the integrated provision of environmental infrastructure;
- promoting sustainable energy and transport systems in human settlements;
- promoting human settlement planning and management in disaster-prone areas; and

● promoting sustainable construction industry activities.

These primarily urban priorities are complemented by a chapter on the need for integrated planning and management of land resources. This aims for the goal of 'integrating environment and development in decision-making'.

The inescapable importance of urban areas is reflected in the fact that:

> By the turn of the century, the majority of the world's population will be living in cities. While urban settlements, particularly in developing countries, are showing many of the symptoms of the global environment and development crisis, they nevertheless generate 60 per cent of gross national product and, if properly managed, can develop the capacity to sustain their productivity, improve the living conditions of their residents and manage natural resources in a sustainable way.
>
> (UNCED, Chapter 7, section B).

Thus, an important objective is to:

> . . . provide for the land requirements of human settlement development through environmentally sound physical planning and land use so as to ensure access to land to all households and, where appropriate, the encouragement of communally and collectively owned and managed land. Particular attention should be paid to the needs of women and indigenous people for economic and cultural reasons.

Equally, in the rural context, it is argued that:

> Land resources are used for a variety of purposes which interact and may compete with one another; therefore it is desirable to plan and manage all uses in an integrated manner. Integration should take place at two levels, considering, on the one hand, all environmental, social and economic factors . . . and, on the other, all environmental and resource components together . . . . Integrated consideration facilitates appropriate choices and trade-offs, thus maximising sustainable productivity and use.

These specific requirements for urban and rural areas must be viewed in terms of Agenda 21's perceived need for integration of environment and development considerations in the broader decision-making process. Thus, it is noted that:

> Prevailing systems for decision-making in many countries tend to separate economic, social and environmental factors at the policy, planning and management levels. This influences the actions of all groups in society, including Governments, industry and individuals, and has important implications for the efficiency and sustainability of development . . . . An overall objective is to improve or restructure the decision-making process so that consideration of socio-economic and environmental issues is fully integrated and a broader range of public participation assured.

It is worth dwelling on the broader perception of the role of and need for land-use planning, before considering an individual land-use planning system. This is because we tend both to become over-familiar with a particular system, not recognising its idiosyncrasies and limitations, and also periodically to lose sight of why planning is necessary in principle. Despite the limitations of 'command and control' approaches to environmental regulation, planning is still widely perceived as an essential link in the chain between global and local sustainability.

Some commentators have proposed that planning is a superseded activity and that a more effective and sensitive use of the market will replace the need for bureaucratic intervention by planners. This line of argument suggests that, given a sufficient appreciation of the real value of environmental goods and the availability of reasonable models of human response to economic signals, the market can be made to deliver the most efficient balance between conservation and development. Thus, one perspective is that planning is an outmoded, collectivist, interventionist and bureaucratic approach which has been characterised by expensive blunders at the implementation stage.

This viewpoint is controversial, but it has contributed to a certain re-definition of the domain and purpose of land-use planning. We must, therefore, be careful to consider what planning could, or should attempt to, deliver in the context of sustainable development. One reason for its continuing importance is its association with place. Thus, whilst economic analysis relates mainly to abstract space, real development and land use take place in precisely located geographical space with all its peculiarities of cultural attachment, landownership, environmental constraints and site history. A locally-sensitive administrative and political framework is necessary to address this type of contingency. Another reason is the practical impossibility of addressing all the externalities, or neighbourhood effects, associated with the use of a particular site. Whilst the market can reflect the principal values of an environmental resource, we do tend to experience 'market failure' in relation to the complex, aggregate, temporal costs and benefits associated with its usage. Thus, there will always be a range of externalities which persist at a particular place over a given time period and which are inefficiently resolved by the market, so that a more interventionist style of arbitration is necessary. A further instance of practical 'market failure' is that, in order to encourage people to adopt more environmentally sensitive practices such as switching to public transport or improving the energy efficiency of their dwellings, market-based solutions may need to introduce swingeing and politically unacceptable price rises. In practice, it may be more feasible simply to introduce physical measures and legal controls to achieve the desired results.

The previous extracts from Agenda 21 also emphasise the value of planning mechanisms as means of integrating environmental and developmental interests, and of striking the right balance between users of a limited resource. This integrative or balancing function requires a system in which trained arbiters can weigh evidence and then intervene. Similarly, it is clear that Agenda 21 sees participative approaches as being essential to sustainable development. Whilst the market does allow individuals to exercise some choice, this is largely manifest as consumerism and 'passive citizenship'. The conditions for active citizenship, and contributing to a consensus-building discourse, can only be met where there is some kind of local planning system. The interests of sustainable development will generally best be served within political systems which have a healthy respect for the co-existence of market-based and interventionist mechanisms, rather than advocating one at the expense of the other.

Sustainability and the changing purpose of planning

The genesis of town planning lay in the desire to reconcile economic prosperity with acceptable standards of living and human development. Many commentators have alluded to the distinctly 'ecological' spirit of early town planning pioneers. These, it is argued, aimed to establish a profession which respected biological limitations, both in human and non-human terms, and which sought to balance the best aspects of town and country to the benefit of all citizens. Early planners thus championed the defence of biophysical systems as well as the promotion of 'model settlements', such as new towns and garden cities. Some would thus argue that the profession has always been concerned for the protection of ecological, or at least human ecological, resources: hence, sustainable development is merely the continuation of a deeply-rooted tradition. Indeed, it could be argued that town planning encapsulates the core concepts of sustainable development. Ideas of ensuring the quality of environment, of handing this intact to future generations (futurity), fairness in the allocation of resources (equity) and right to be heard in development choices (participation) are integral parts of the town planning heritage.

However, whilst some planners have sought to claim the high ground of sustainability for the planning profession, it must be recognised that the pressure for 'green planning' has not come from 'green planners' as such. The primary impulse has been the eventual acceptance, at the very highest levels of government, of the validity of environmentalists' claims about potentially irreversible deterioration of natural capital stocks. In their concern to find rapid ways of preparing and delivering sustainability programmes, governments have had to rely heavily on the capacity of existing professions. Consequently, one interpretation is that the planning system has virtually had 'greatness thrust upon it' as a key player in national and local sustainability strategies.

Notwithstanding the fact that some critics have viewed the UK's NSDS as 'light green', and superficial in its reappraisal of economy-environment interrelationships, it has given welcome and clear affirmation of the need for long-term decision-making and the planning of land use investment decisions. The NSDS is one of the most positive recent affirmations of the value of planning in respect of land development, built environment, minerals and countryside. For much of the 1980s, planning occupied a politically subservient position, being equated with old-fashioned 'statism'. Some politicians also disliked its concern for subjective matters of taste and value judgement in relation to design. Since the early 1990s, this has changed: not only are aesthetic matters considered relevant to planning decisions, but the broader environmental agenda is deemed to enhance the essential spirit and purpose of planning.

Complementary to this changing perception has been the emergence of a European perspective on land use planning. The fifth EU Environmental

Table 6.1   FoE view of development plans' role

Local planning authorities should promote an environmentally sustainble quantity, pattern and form of development, through which the land use pattern will facilitate the conservation of energy and other natural resources and the minimisation of pollution, particularly through the reduction of the need for travel. The authority should welcome new development where it is most appropriate and of most benefit to local and regional needs, provided that it does not have a detrimental impact on local amenity, the environment and the means whereby future needs will be met. The spirit and content of this policy should cascade down through the other policies in the local plan. Councils should develop policies in the areas of:

- *energy* supply and use, including the pursuit of renewable energy sources, combined heat and power, energy conservation, energy implications of changes in travel patterns, and radioactive waste;

- *transport*, including significant travel restraint (aiming for a 30% reduction in traffic volumes by 2005), traffic calming, encouraging cycling and walking, and using land use allocation to reduce the need to travel;
- *waste management* practices in the areas of waste generation, disposal, reuse and recycling facilities, landfill, sources of specific waste, and transport of specific wastes;
- *habitat protection and greenspace* protection and management for wildlife sites, open space, allotments, and wider rural land uses; and
- *water and air pollution* including protection of water abstraction points and aquifers, water use and conservation, levels of atmospheric pollution, and emission sources.

To complement these activities Councils should develop:

- *planning tools and procedures* such as environmental assessment, planning obligations and planning standards.

Source: based on Bosworth, 1993

Action Programme, *Towards Sustainability*, acknowledges the need for 'integrated sectoral and spatial planning, whilst *Europe 2000* advocates EU-wide approaches to the planning of settlements and transport systems (CEC, 1990a,b). If a more homogenised EU planning system is to emerge, then it is likely that sustainability will be one of its cornerstones.

Blowers (1993a) has similarly evaluated the essential features of town planning which correspond to the canons of sustainability. First, he notes that planners must analyse the options which can be taken in relation to an uncertain future. This, in principle, is analogous to the use of a 'precautionary approach'. Second, planning is an integrative activity, and thus can help tackle the integrated and multi-disciplinary nature of environmental processes and policies. Third, planners must take a strategic view of the way that decisions are made. This not only refers to timescales, but also to the decision-making processes at national, intermediate and local tiers, and to the strategic and tactical bases of policy development.

On a more pragmatic plane, Bosworth (1993) considers that planners' ability to respond to sustainability issues has been enhanced during the early 1990s. Partly, this arises from the district-wide coverage of local plans, which requires that the whole countryside is covered, rather than the traditional hot-spots of development pressure. Also, since plans now have to take account of the environment 'in the widest sense', they no longer dwell solely on former narrowly defined environmental issues of housing, traffic, land use and amenity. Finally, since plans have become the focus for individual planning decisions, their long-term and integrative analysis of environmental needs lies at the centre of development rather than at the periphery. Bosworth argues that local planning authorities should have policies on energy, transport, waste management, habitat protection, water and air pollution, and should develop the planning tools and procedures to address these topics (Table 6.1).

The agenda for 'sustainability planning'

Not only does town planning embrace sustainability within its wider spirit and purpose, but also in its detailed agenda. Even its traditional domain of amenity and physical land use contains topics of obvious relevance to modern environmentalism. These include the siting and design of buildings, the restriction of urban sprawl, and the (partial) safeguard of important areas of landscape. Until recently, though, many other sustainability topics fell outside the explicit remit of planning as material considerations. Now, there is a far greater willingness, even encouragement, from government to embrace a wider range of 'green' issues.

This more recent agenda includes the reduction in consumption of natural resources, such as lowering CO_2 emissions by reduced energy use and encouraging recycling. The safeguard and wise use of land is obviously at the heart of town planning, and has been stressed in, for instance, the White Paper on the countryside (DoE/MAFF, 1995). Much greater emphasis is starting to be placed on the recycling of 'brownfield' sites, especially contaminated or under-used land, and resisting the development of greenfield sites. This is coupled to firm green belt policies and to the resistance of urban sprawl and protection of open land. Green belts are often perceived as a negative policy instrument of restraint but they can also be seen positively aimed at regenerating urban areas and providing recreational and environmental amenities for the city. Physical environmental protection can also be interpreted as covering far more than aesthetics. It may embrace the entire cultural heritage stock of town and country, including areas of high landscape and wildlife value as well as areas of more accessible open space, which are important both for human quality of life and for the functioning of natural cycles within urban areas. Equally, 'good design' is about far more than 'taste', and can result in buildings being more efficiently related to their natural surroundings and environmental conditions.

A significant area of environmental planning is now concerned with the conservation of energy and reduction of pollution. This involves discouraging development in locations which would exacerbate congestion, as well as planning at a more strategic level to integrate transport and land use decisions. Whilst restricting car use is extremely difficult, given that cars are so convenient and even necessary for people with complex time budgets, it is still important to consider the location of development so that it minimises the need for journeys, or at least makes them more feasible by public transport. Thus, trip generation, associated with workplace and retailing activities, can be encouraged at nodes in the public transport system. Some local authorities have, for example, promoted 'sustainable' offices and industrial units in locations well served by bus routes. Similarly, town centres are worth maintaining, not only for their inherent cultural value but also because they are suitable for *combining* trips for different purposes, such as shopping, leisure and employment.

Complementary strategies are also possible, such as improving neighbourhood shopping centres so that the need for longer distance trips is reduced, promoting a balanced (self-contained) structure of land uses in certain districts in order to mix residential and 'neighbourly' employment uses, encouraging walking and cycling and facilitating energy efficiency through energy conservation and combined heat and power schemes. One of the most contentious debates in countryside planning at the moment is the role of 'new' settlements, that is, relatively free-standing residential communities in the open countryside as an alternative to constant accretion to existing towns and villages. On the benefit side, they may offer scope for innovation in the efficient use of energy and water resources and may also be located close to rapid public transport routes. However, they can have considerable environmental impact, threaten green belt and 'brownfield' development policies and they can undermine the regeneration of some rural communities. Provided they are conceived so as not to induce undesirable pressures for decentralisation, it seems that new rural settlements are likely to gain a place in the future countryside, and must thus be designed as environmental showpieces.

These considerations are all essentially *products* which the planning system can help to deliver. Perhaps just as important is the *process* of reaching decisions, which it can also aid. Thus, development plans must now be subject to a process of explicit environmental appraisal, and this will include a wide process of consultation and consensus-seeking. A reasonably wide range of development applications must similarly be subject to environmental assessment, and some local authorities have widened the net to catch far more 'normal' development control applications within an informal EA system. Thus, the public is given a far greater opportunity of contributing to the environmental debate. This is important, as sustainable development will entail building consensus in essentially problematic and disputed subject domains. For example, people may strongly prefer low-density living and high personal mobility, whereas these preferences may be in conflict with environmental goals of prudent resource use. These issues must not be side-stepped, but, as with all 'wicked' problems must be debated and reconciled within the context of a formal democratic decision-making framework.

Planning policy

It is evident that central government now sees the planning system as a key instrument in the delivery of sustainable development. A major means of influencing local planning practice in this direction is through its guidance on key policy issues. Within England and Wales, this has mainly been effected through Planning Policy Guidance (PPGs) and Mineral Planning Guidance Notes and in Scotland through National Planning Policy Guidelines. In England and Wales, PPGs now cover over twenty topics, the majority of which are of direct or indirect natural environmental significance and many of which were revised from 1992 onwards to reflect sustainability principles. The key PPG in this respect is Number 12, which emphasises that:

> . . . the Government has made clear its intention to work towards ensuring that development and growth are sustainable. The planning system, and the preparation of development plans in particular, can contribute to the objectives of ensuring that development and growth are sustainable.
>
> (DoE, 1992).

In addition to its general tenor, the Guidance Note includes advice on specific topics such as global warming and land recycling (Fairlamb, 1994). Other relevant PPGs are those on general policy and principles (No. 1), housing (No. 3), industrial and commercial development (No. 4), town centres and retail development (No. 6), countryside and the rural economy (No. 7), nature conservation (No. 9), transport (No. 13), sport and recreation (No. 17), renewable energy (No. 22) and planning and pollution (No. 23). Although these are perhaps conservative in terms of their association with traditional areas of environmental concern (such as protection and amenity), they also include challenging statements about efficient built form, transport and service provision (Table, 6.2).

Central policy guidance is a mixed blessing in that, whilst it provides some elegant justifications for robust and often innovative planning approaches, it precludes a certain amount of local independence and nonconformity. It does, however, help reorient policy direction and signal the kind of interpretation which the Secretary of State and the Planning Inspectorate are seeking to place on individual decisions. Sometimes it appears vague in relation to the 'wicked problems' of sustainable development which present themselves locally, and its interpretation is not always unambiguous. Nevertheless, PPGs have been extensively reviewed to take into account sustainability principles, and are valuable allies in planners' defence of environmental quality.

Table 6.2 Elements of Planning Policy Guidance Notes reflecting sustainability principles

---

PPG1 (March 1992) General Policy and Principles
This PPG sets down the general principles and functions of the planning system. It was first released in 1980 but subsequent revisions (e.g. in 1992) conceded a more affirmative role to planning. It now includes explicit statements about government commitments to work towards ensuring that development and growth are sustainable.

PPG3 (March 1992) Housing
This relates government housing policy to planning and, in the context of sustainability, notes:

- the increased emphasis on the reuse of urban land as a means of relieving pressures on the countryside;
- the emphasis on ensuring that new housing development does not damage the character and amenity of established residential areas; and
- the need for authorities to make appropriate provision for affordable housing.

PPG4 (1992) Industrial and Commercial Development
Revised in 1992, this Note placed increased emphasis on environmental considerations, whilst meeting the needs of industry. Specific reference is made to the need to control the emissions of greenhouse gases. New themes included:

- encouraging new industrial development in locations which will minimise the length and number of trips;
- encouraging development in locations that can be served by more energy efficient modes of transport; and
- discouraging development where it would be likely to add unacceptably to congestion.

PPG6 (1993) Town Centres and Retail Development
This goes some way towards seeking to ensure that new out of town developments should only be allowed if they do not threaten the viability of town centres. It considers:
- the need for retail development to be located where it will be accessible to all sectors of society and provide a choice of transport modes;
- the role of town centres in minimising travel demands, especially by car;
- the role of neighbourhood centres and local shops in reducing the need for people to travel to towns for all their needs; and
- the need for good town centre management and a positive policy approach to town centres.

PPG7 (1992) The Countryside and the Rural Economy
Whilst seeking protection of the countryside, this PPG aims to encourage economic activity in rural areas to offset the decline in farming incomes. The diversification of the rural economy is to be encouraged, as is the re-use of redundant agricultural buildings. An important element is to seek the protection of countryside for its own sake, that is, the more integrative and indefinable qualities of rural amenity and quality of life, rather than predominantly agricultural values.

PPG 9 (1994) Nature Conservation
This outlines ways in which planning authorities can assist the pursuit of national and international obligations on nature conservation, as well as supporting local initiatives for conservation provision. It also indicates ways in which safeguard and promotion of wildlife sites and corridors in the wider countryside can contribute to the viability of nature.

PPG12 (revised 1992) Development Plans and Regional Policy Guidance
This guidance is of major significance in respect of policy towards the environment, especially the formulation and content of development plans. Whilst noting that traditional environmental concerns remain valid, it stresses the importance of development plans in contributing to the objectives of sustainable growth. It stresses that environmental issues are an integral consideration which must be applied to the full range of development issues including housing, transport, energy generation and the economy. A specific requirement for the environmental appraisal of development plans is included to ensure that the environmental costs and benefits of policy are systematically assessed. A number of 'sustainability' themes are contained, namely:

- recognition that land is a finite resource;
- the need to sustain the character and diversity of the coast, countryside and wildlife resource;
- the importance of green belts in resisting sprawl;
- the need to maintain the character and vitality of town centres;
- the importance of safeguarding the amenity and character of established residential areas;
- the importance of good design and protection of the built heritage;
- the relevance of location policy in reducing the need to travel (and thus conserve energy and reduce CO_2 emissions);
- the importance of public transport; and
- encouraging the use of vacant, derelict and contaminated urban land.

PPG13 (revised 1994) Transport
This recognises the key role of transport and achieving sustainable development. Its two key themes comprise:

- the desirability of reducing the need to travel (and reliance on the car); and
- the need better to integrate land use planning and transport provision.

It is recognised that the key areas where planning has a role to play relate to density, settlement size and structure, town centres, neighbourhood planning and the integration of land use and transport planning. Advice is given on location policy as it refers to a range of land uses and wider issues, including:

- *housing* – maximise growth in existing large urban areas, promote development in sites well served by public transport, avoid incremental expansion and dispersed settlement patterns, juxtapose employment and housing where appropriate;

- *employment* – locate offices centrally, increase self-containment in urban areas where possible, avoid large-scale and high intensity development in areas poorly served by public transport;

- *retail/leisure* – maintain and revitalise existing town and suburban centres, promote local shopping facilities, favour edge of centre superstores rather than out of centre development, concentrate leisure activities in town centres;

- *neighbourhood planning* – since a high proportion of transport is local, planners should provide safe and convenient pedestrian/cycle access, seek to secure a range of facilities and services in local centres, allocate residential development near existing centres, encourage walking and cycling, and deter car use in town centres; and

- *integration of land use and transport planning* – regional guidance should fully examine the interactions between land use and transport infrastructure, local accessibility profiles should be used to assist policy development, and local authorities should protect and maintain the competitiveness of town centres.

PPG17 (1991) Sport and Recreation
Planning authorities are urged to take full account of a community's need for recreational space, to have regard to current levels of provision and deficiencies and to resist pressures for the development of open space which conflict with the wider public interest. The guidance also contains a strong defence of the recreational and amenity value of open space, and its contribution to the quality of urban life.

PPG 22 (1993 + 1994 supplement) Renewable Energy
This PPG emphasises the government's commitment to developing renewable energy sources (e.g. energy from wind, sun, water, etc.). It acknowledges the role of renewable energy in reducing the supply of harmful emissions, and helping the UK meet its targets for reduction of greenhouse gases. In preparing development plans, planning authorities should take into account the benefits of renewable energy and consider the contribution which their area could make to its development, bearing in mind other relevant considerations. Various technical annexes are included on the various forms of renewable energy: wind energy, waste combustion, hydro power, anaerobic digestion, landfill gas, solar systems. The PPG considers various issues relevant to planning, namely, the technologies involved, licensing implications, siting issues, visual effects, traffic and other local environmental impacts.

PPG23 Planning and Pollution
This PPG emphasises that pollution issues are normally the matter of other agencies (now, chiefly, the Environment Agency); the aim of the PPG is thus to encourage close collaboration between planners and other environmental professions, rather than to duplicate controls. The PPG explains the operation of *integrated pollution control* and the interrelationship of waste management planning and guidance frameworks. Given that pollution control lies mainly outside the planning system, the PPG emphasises those matters which do represent material planning considerations. These may include the availability of land for potentially polluting development, the sensitivity of the area, loss of amenity, and any practical environmental benefits which may accrue from development, such as regeneration of derelict land, transport improvements and the infill of former mineral workings.

Source: Based partly on Fairlamb, 1994)

Development plans and the environment

In the plan-led system which has been in operation since 1991, development-plan policies clearly have considerable significance in shaping future environments and it is important to be aware of the means by which they may include sustainability matters. First, development plans may include specific 'green' measures such as energy conservation or local nature reserve creation. This is a common approach, yet one in which well-intentioned proposals may be neutralised by contradictory policies and proposals in other sections of the plan. Thus, for instance, a plan may contain policies to protect a local area of estuarine wetland for environmental education purposes, which are elsewhere undermined by permissions to expand polluting industry upstream of the site. Second, a 'light green' approach may involve placing the environment chapter at the front of the plan, laying down strategic markers for subsequent chapters. Whilst this was, until recently, the preferred option for environmentalists, it has rarely exerted the desired influence over more general plan policies.

Despite the fact that environmental management policies may have been strongly stated in a prominent position, this has not in practice prevented subsequent chapters on housing, transport and retailing, for instance, including environmentally insensitive policies. The third, 'deep green', approach entails testing all policies against explicit sutainability criteria. Whilst it is probably impossible for a plan to be entirely supportive of the pursuit of sustainable development, such an approach at least ensures that all trade-offs and unresolved problems are made explicit.

Attempts at deep green approaches are in their infancy, but one example of key strategic objectives which offer criteria for general plan evaluation has been set by Hertfordshire County Council (1994). These criteria allow HCC:

- to enable activities and development to be carried out in the most sustainable way possible;
- to improve the overall quality of life for residents, workers and other users;
- to encourage people to make sustainable choices by ensuring they are also the easiest and most desirable choices;
- to allow the same degree of choice to future users; and
- to contain consumption of, and damage to, natural resources.

These criteria have not satisfied all critics but they are a valuable starting point and they do illustrate the ways in which sustainability criteria can be used as a basis for an overall evaluation of plan policies.

Land-use plans are valuable for reconciling the general issues associated with conservation and development but, at the site level, other approaches will prove necessary. One mechanism which has gained prominence in recent years is that of 'planning gain'. In this, a planning authority seeks to secure a wider environmental benefit from a development proposal, in order to overcome planning objections to the use of a particular site. This typically entails alleviating environmental impact, or creating new environmental assets – in practice, assisting the maintenance or enhancement of environmental capital. In order to secure appropriate environmental benefits, enhancement measures may need to take place outside the actual development site, and thus not legally be susceptible to resolution by planning conditions. To date, planning obligations have had disappointingly little application to sustainable development solutions, but there is evidently some scope for the creation of local landscape features and more strategic environmental projects (e.g. Boucher and Whatmore, 1993). Planning gain of this nature could open up possibilities for compensatory provisions following environmental damage, and would thus assist the maintenance of a constant stock of substitutable natural capital.

Environmental assessment

Environmental assessment (EA) is now a well-established procedure for scrutinising development proposals in terms of their likely effect on the environment and human settlements. In some countries it is a free-standing process and represents the major means of vetting development activities, whilst in others it is embedded into a more mature regulatory framework. In Britain, EA is mainly exercised through the town and country planning system, though certain proposals (e.g. forestry, highways, land drainage) are administered under other legislation.

Very broadly, the EA process involves (after Glasson et al, 1994):

- project screening – narrowing the application of EA to those projects which have significant impacts;
- scoping – narrowing the focus of EA to the principal issues of significance associated with a project;
- consideration of alternatives to the project proposed;
- description of the characteristics of the project, in terms of its construction, location and processes;
- identification of key impacts, both adverse and beneficial;
- prediction of impacts, in terms of their likely magnitude;
- evaluation and assessment in terms of the relative significance of predicted impacts, permitting assessors to focus on key adverse impacts;
- presentation of the overall statement of impacts;
- review of the quality of the final statement;
- decision-making, balancing the information in the environmental statement with other material considerations;
- post-decision monitoring of the development, if it proceeds, contributing to effective project management; and
- auditing of actual outcomes with predicted outcomes, to help refine the art and science of environmental impact forecasting.

The main documentation associated with this process is referred to as the environmental statement (ES) and typically contains four main sections, though this practice is by no means universal. First, a *non-technical summary* presents complex environmental issues in lay terms and serves to aid communication with various interested parties. Second, a *methods statement* sets out at an early stage the personnel involved, the approach taken and perhaps some of the problems they encountered. Third, a section covering the *background to the proposed development* reviews the nature of the site and prevailing (baseline) environmental conditions and describes the proposed development and its construction. The last major component is the main body of the statement which comprises a *topic area* coverage, reviewing significant impacts in key areas. These typically include:

- land use, landscape and visual quality;
- geology, topography and soils;
- hydrology and water quality;
- air quality and climate;
- terrestrial and aquatic ecology;
- noise;

- transport;
- socio-economic; and
- interrelationships between effects.

Quite extensive guidance on the appropriate methods and techniques to be used are now available in many countries (e.g. DoE, 1995). The purpose of the ES should not be solely to provide expert advice on the environmental management of a project, but should also inform and involve the affected public and other interested parties.

In practice, this idealised procedure has not always been followed, and case histories point to various sources of controversy. For example, the range of projects potentially subject to EA is quite narrowly circumscribed by legislators, but best practice appears to be widening the type of projects suitable for appraisal and the spectrum of topics considered. Furthermore, some of the key steps outlined in the ideal process may commonly be missing, with monitoring and auditing provisions particularly poorly addressed. It has also been noted that the balance of influence between the various parties involved in the EA process is inconsistent, particularly in the degree to which emphasis is placed on public consultation and participation. Moreover, the quality of ESs has been highly variable, with many failing to meet even the minimum standards.

To date, EA has dealt mainly with development projects, that is, specific applications to develop particular sites or routeways. This approach is limited in terms of achieving sustainable development because it hearkens back to a period when environmentalists had to react to development proposals and engage in damage-limitation exercises, rather than anticipate change and ensure that is was sustainable in terms of environmental potentials and capacities. The latter can only be attempted (and even then only with great difficulty) if policies, plans and programmes are assessed for their environmental effects so that greater attention is directed towards the preferred types and locations of development before they reach the 'firm proposal' stage. Methods of strategic environmental appraisal are still in their infancy, and progress is slow because of the inherent difficulty of anticipating future change with any confidence.

Guidance from DoE (1993b) on the strategic appraisal of development plans states that it should be:

- an explicit, systematic and iterative review of development plan policies and proposals to evaluate their individual and combined impacts on the environment;
- an integral part of the plan-making and review process, which allows for the evaluation of alternatives; and
- based on a quantifiable baseline of environmental quality.

This cannot happen if EA is simply seen as an add-on, and it is only likely to work if it forms an integral part of the process of drawing up a plan. The process broadly seeks to:

- 'characterise' the environment, looking at key assets, threats and opportunities in order to provide a 'baseline' and context for considering the environmental effects of policies;
- ensure that the scope of the plan covers the appropriate range of environmental concerns; and
- appraise policies and proposals to estimate their environmental effects.

The environment is appraised in terms of its 'stock' of human and natural resources; and the plan is then 'scoped' to assure that it includes the requisite coverage of issues, which may typically be inferred from central government guidance. The recommended approach is then to appraise plan content in terms of its internal consistency, so that development proposals accord with its overall environmental objectives and that individual policies do not inadvertently cancel out the environmental targets of other policies. A popular way of checking this is through a compatibility matrix, in which elements of environmental stock are arranged along one axis and policies on the other, to facilitate an explicit exploration of the impact of each policy option on each aspect of environmental stock. The matrix is used to record whether there is likely to be a positive (enhancing), negative (harmful) or neutral impact.

People and planning

The need to include local views in the preparation of plans, and in selective consultations over specific development proposals, has been accepted for a long time. Participation by the public in seen to increase the legitimacy of the planning process, lead to the production of better informed decisions, and raise public interest in planning matters. However, methods of encouraging public viewpoints have often been rather unimaginative. In terms of Arnstein's (1969) celebrated 'ladder of participation' (Figure 6.1), they have rarely been more than consultative in nature. Whilst Arnstein's schema may be open to question, it does suggest that there is scope for far more invention if local sustainability plans are to take citizen involvement (or even empowerment) truly seriously.

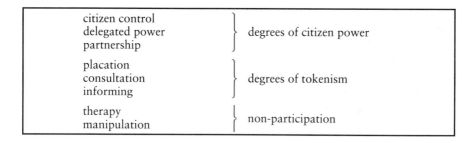

Figure 6.1   A ladder of citizen participation
Source: based on Arnstein, 1969

Several studies emphasise the importance of the locality as an arena for effective citizen action (for a recent review, see Aygeman and Evans, 1994). One line of argument is that social responsibity is likely to evolve out of deliberating with others as to what human projects are possible, and it thus requires more opportunities for deliberation above, and particularly below, the nation state (Twine, 1994). This view has obvious relevance to LA21s, with their apparent concern for the decentralisation of decision-making downwards and development of international cooperation upwards. The local state, which many political commentators believe to be distinguished from the central state by its emphasis on consumption and representational functions, may be an intrinsically appropriate focus for attainment of Agenda 21 goals, many of which are related to consumer behaviour and democratisation. It has, further, been argued that it is at the municipal and community level that most of the policies are made that shape and even force many of our current consumption patterns, where vested interests are more manageable and the voice of personal conscience may be heard (Boothroyd et al, 1994).

This places the participatory role of planning in a particularly important position. It is not merely a case of the public being offered an opportunity to contribute to the planning process *per se*, but one of plans becoming central to the forum of debate and communication which is essential for sustainable development. There has been a good deal of argument latterly about the 'communicative' role of plans, and their scope to help people 'reason together' in order to generate a discursive consensus, and thus assist the reconciliation of complex local issues (e.g. Fischer and Forester, 1993). It has previously been noted that a major need in environmental management is for consensus-building and conflict resolution, and the development plan may have a major purpose in this respect. This is particularly so in the case of local sustainability, where the major communicative work of plans takes place at the level closest to the community (local and unitary development plans).

The more progressive work which has taken place on public participation will thus be of major relevance to local sustainability. One of the most effective approaches has been that of 'planning for real' exercises, in which local residents are engaged in facilitative meetings with planners, and typically have model representations of the locality available for demonstration of planning options. The creation of an egalitarian atmosphere, and a more 'concrete' representation of the neighbourhood and its issues have been enormously successful in helping identify better solutions and in giving the affected public a genuine sense of ownership.

The method has been most widely tested to address 'physical' issues of town planning, and its extension to a broader range of environmental topics presents problems, if only because the geographical area is wider and the issues less easily representable. However, the method can be used, with appropriate adaptations. Potter et al (1994), for instance, tested a similar approach in a research context for an area in the urban fringe of Ashford, Kent, using aerial photographs on which people could annotate perceived problems,

Table 6.4 Suggested markers for the evaluation of sustainable development in planning statements

---

Absolute protection of critical natural capital
- defence of air quality (transport, planning and pollution, CO_2 fixing);
- defence of water quality (planning and pollution, commitment to catchment management plans);
- defence of key habitats (planning and nature conservation, coastal zone planning); and
- minimum depletion of non-renewable resources (mineral planning, recycling/reuse, renewable energy, energy efficient buildings).

Maintenance of constant stock of substitutable natural capital
- use of planning obligations for natural resource compensation;
- maximum use of restoration and reclamation powers;
- constantly ensuring viability of, or finding new uses for, traditional employment/retail/residential zones; and
- involvement in strategies for new woodlands/forests.

Futurity of decision
- long-term planning horizons;
- full values of natural capital to be represented in decision-making techniques;
- use of appropriate discount rates and horizons in decision-making techniques; and
- use of strategic and policy impact assessment.

Inter-society and inter-generation equity
- consideration of ecological footprints on areas outside municipality or local authority area (including overseas);
- local authority commitment to overseas links;
- constant pursuit of 'quality of environment' objectives;
- involvement in European 'urban environment' initiatives; and
- local authority green audits to consider product origins and destinations of wastes.

Virtuous development circle
- optimum use of environmental assessment procedures in project appraisal;
- use of policy consistency analysis in development planning;
- use of environmental planning techniques based on capacities and thresholds;
- availability of high quality environmental information to decision-makers; and
- good example set by local authority on energy, materials and transport policies.

Encouragement of citizen views and actions
- transparent decision-making;
- imaginative role for elected members;
- creation of representative local forums;
- recognition of principle of subsidiarity; and
- development plans which optimise their 'communication' function.

Adoption of a robust process leading to widely 'owned' and reliable products
- local authority green auditing procedures; and
- local authority 'exit strategies' for appropriate, self-sustaining initiatives; and
- procedures for building consensus and mediating conflict.

Source: Selman, 1995

opportunities and possible solutions. In a practical application, the Brecon Beacons National Park Authority used the approach to help them prepare their first park-wide local plan (Thomas and Tewdwr-Jones, 1995). The planners considered that they 'needed some way of getting to the heart of complaints, to disperse the confrontational atmosphere, to encourage consensus and involve the community in the decision-making process'. Their approach was to use detailed GIS maps rather than models because it was a large, upland, sparsely-populated area. The format of the meetings, in terms of room-layout and activity, was that of a 'structured jumble sale' in which easy-to-read maps were used to provide a central focus, and posters posed questions aimed at generating ideas and debate. The public were then able to use coloured pins, flags and model houses to demonstrate where they thought there were significant issues, such as traffic problems, housing needs and important open space. A series of thirty-nine meetings was held, the outcome of each one being a list of specific issues, which planners could then use as a focus for writing the local plan. Several problems arose which were not 'planning matters' but, rather than ignoring these, 'action sheets' were designed which could be passed to the appropriate agency or department.

Conclusion

The role of planning in local sustainability has been widely attested. It has the capacity to take into account the main principles of sustainability, such as futurity, the human and natural environment, equity, and local community participation. Local plans are now becoming geographically comprehensive in their coverage, they may be environment-led, and are required to take account of the environment 'in the widest sense'.

Sustainability has given renewed vigour and justification to many of the most basic tenets of planning, such as protection of prime resources, local relevance and demand management. Many recently-adopted development plans may be dressed up in a fashionable new jargon, but their basic policy thrust is entirely consistent with the most time-honoured principles of wise stewardship. Moreover, governments are now unavoidably committed (through international protocols and treaties) to many long-term environmental measures, and planning legislation is one of the sturdiest straws at which they can clutch. Sustainable development brings with it, however, abundant scope for jargon and empty rhetoric. For planning to make a genuine contribution, it must be systematically checked against sustainability criteria; a selection of contenders for these criteria is offered in Table 6.4.

7

Businesses and local sustainability

Introduction

Central to the achievement of sustainable development is the reconciliation of economic and environmental objectives. For ease of discussion, industry and commerce are rather inaccurately referred to as 'business'. Although the early environmental movement saw business as the villain of the piece, there is now a generally acknowledged need to consider the 'competitive environment' alongside the biophysical environment. Thus, Agenda 21 sought to:

- encourage the concept of stewardship in the management and utilisation of natural resources by entrepreneurs; and
- increase the number of entrepreneurs engaged in enterprises that subscribe to and implement sustainable development policies.

These are ambitious intentions, strongly constrained by the economic realities within which the business community acts. Clearly, it is essential to gain the genuine and practical commitment of businesses to sustainable development, but it is also difficult to convince them to take anything more than a symbolic gesture to reducing their environmental impact. Moreover, local businesses are frequently branches of national and transnational enterprises and thus may not be free to respond independently to 'green' initiatives, particularly if their response necessarily entails adopting new corporate environmental mission statements and internal auditing procedures.

It is therefore important not to be glib about the likely contribution of business to sustainable development, nor to focus on isolated instances of sponsorship or showpieces. The pessimistic view is that, despite the rhetoric of the business sector, production is driven by consumption, deriving from market forces and advertising, and this leads inexorably to profligate and unsustainable behaviour. Fortunately, there is an altogether more optimistic view of green business. This supposes that environmentally-sensitive business is more efficient, because it creates fewer externalities (such as pollution and congestion) and enjoys increased respect from consumers and employees. There is, indeed, some evidence that consumers prefer products which are 'environmentally friendly', or at least which have caused minimal adverse environmental impact on a cradle-to-grave basis. This is reinforced by pressure from the media, the community, employees and shareholders for environmentally responsible behaviour, and for 'chaining' this responsibility through contracts to suppliers and contractors.

There is little in the literature which relates specifically to the role of *local* business in the sustainability process, and so much of this chapter looks at the wider environment within which business operates. However, corporate business is important to local sustainability for a number of reasons. Four evident examples of this contribution are that:

- corporate business has been the pioneer of a suite of methods and techniques relevant to local sustainability – notably those relating to environmental management systems, consensus-building, environmental accounting and 'envisioning' – and so it can often provide valuable advice, exemplary leadership and expertise to Local Agenda 21;
- business contributes significantly to local environmental initiatives through sponsorship, though not perhaps as dramatically as some would like; however, businesses are often more able to contribute free skills (such as the sustainability techniques mentioned above): contributions made in kind may be more useful than cash;
- the workplace is a base for adult education, in an informal sense, and thus may contribute to the 'greening' of employees; and
- local businesses may occupy critical, intermediate positions in relation to corporate greening, either because they are branches of larger corporations which possess centralised environmental policies, or because they form a link in the supply chain to larger companies who require compliance with their own corporate green policies.

Thus, business can be important in providing environmental leadership and support to the local community. One example of the way in which this role may be formalised is illustrated by the Centres of Environmental Excellence set up by Lancashire County Council for diffusion of best practice to local businesses. Gibbs (1993) has also reviewed the importance of local government itself in stimulating business investment, in relation to wider environmental policy.

The world of 'green' business

Local sustainability at the level of private enterprises must be viewed in the wider context of the general greening of business. This reflects both 'enlightened self-interest' in marketing terms and a growing recognition of the efficiency of reducing environmental impacts. Thus, Clark et al (1993) have argued that:

... any business will be more sustainable if it: makes more durable products; simplifies production processes; uses resources (especially energy) more efficiently; reduces transport; uses reclaimed and renewable resources instead of finite ones; minimises wastes and recycles what it can.

Broadly, there have been two types of positive response from industry to 'green' issues. One has been to cater directly for green consumerism, by producing goods marketed for their environmental friendliness, ranging from detergents to holiday villages. The second has been for companies to ensure that, whilst not a primary marketing proposition, the achievement of reasonable standards of environmental sensitivity is inherent in their product ranges and that their products display environmental information on their packaging. Both of these probably represent only partial and transitional solutions. The former carries the risk that companies which restrict the provision of environmentally sound goods to expensive 'niche marketing' are likely to be caught out by more general shifts in consumer behaviour (Clark et al, 1993).

Demand for certain types of explicitly 'green' goods has proved disappointing in Britain, but of more general significance is the apparent ground swell of expectation that business will behave responsibly towards the environment. It has been argued that businesses which are merely minimalist in their environmental performance targets risk encountering consumer resistance, or the introduction of environmental regulations which they had not anticipated. Not surprisingly, it has been widely suggested that business needs a serious commitment to the environment at a corporate level if it is to cater for consumer demands and be proactive in relation to legislative changes. Dyllick (1989, cited in Hopfenbeck, 1993), argues that successful businesses will not so much be those which specifically seek to exploit 'green' markets, such as pollution control technology, but those manufacturers who are devoted to competitive green production methods and products.

By the same token, businesses cannot all aspire to the same levels of environmental performance, nor to adopt the same type of environmental management practices, as they start from radically different base-lines in terms of their operational characteristics. Roberts (1995) identifies some of the factors which may adversely and (to an extent) unavoidably influence the inherent 'greenness' of a company. These include:

- a high level of dependency upon raw materials that are obtained from mining operations or from sensitive ecological habitats;
- a location in an area of natural beauty or near to a major centre of population;
- a high degree of dependence on long-distance road haulage for the supply of materials and the distribution of products;
- the need to consume large volumes of water;
- an energy-intensive method of production;
- inherently hazardous or environmentally disruptive production processes;
- an inability to transport its workforce by means other than the private car;

- difficulties experienced in developing or utilising methods for the recycling of waste; and
- problems related to the safe disposal of waste.

These problematic environmental circumstances also typically cluster, either within particular industrial sectors or parts of the country, so that regional and sectoral ('structural') factors will further affect environmental performance. Environmentalists cannot expect the overnight transformation of regionally or structurally disadvantaged businesses, but equally there have been striking examples of such companies moving progressively towards sustainability.

A further way in which businesses can seek to green their activities is in the 'chaining' of responsibility, so that their sub-contractors and suppliers have to satisfy environmental criteria. Many major companies are becoming increasingly demanding in this respect, if only out of concern for damage to their own reputation. This cascading of green responsibility is especially important in a country like Britain, where large retail groups wield enormous power. Companies which supply these major retailers thus find that they fail to win contracts if they lack credible environmental management systems. For example, B&Q (1995) set out in 1990 to ensure that, by 1994, all their suppliers should be able to demonstrate a detailed awareness of the pertinent environmental issues and be committed to reducing their impacts. By this target date over ninety per cent had achieved compliance and, during the following year, the remainder either complied or ceased to supply the company. Having reached this baseline, B&Q then sought to pursue real improvements, and introduced a scheme which assessed all their products in terms of Quality, Ethics, Safety and Treatment (QUEST) criteria. In effect, this scheme was an integration of the product performance and environmental aspects of quality assurance. Businesses in their supply chain are now measured on ten QUEST principles and are graded on published policy, actions, involvement in international supply chains, and packaging, as well as more conventional aspects of product quality performance.

The business sector has been increasingly attracted to the concept of total quality management, which is concerned with the implementation of processes and methods designed to eliminate errors and to seek continuous improvements in product performance. This discipline of quality assurance is readily transferable to environmental management systems, where continuous improvement of environmental performance is the objective. Just as TQM in the production of goods is directed towards 'zero defects', so its application in the greening of business has become concerned with 'zero negative impact on the environment'. Thus, the reduction of pollution or of the use of a scarce mineral, for example, is an insufficient end-point: the ultimate objective is to achieve continuous improvement, for instance of production processes, so that there is no adverse impact at all in the long term.

Broadly speaking, businesses will be able to target action in seven main areas, namely:

- use of resources;
- nature of production processes;
- transport operations;
- recycling and reuse;
- waste management and disposal;
- energy use; and
- environmental awareness training and relationships with the local community.

The last of these, in addition to focusing on specific opportunities for 'greening' within the business, provides a major venue for more general employee education. After leaving school, adults can often most effectively be reached via their workplace, and this takes on a key role in developing environmental citizenship and raising awareness. Overall, therefore, business can make a highly positive impact on sustainability, and this collective potential has been encapsulated in a Charter produced by the International Chamber of Commerce (1991) (Table 7.1).

Green business organisation

The operation of all businesses is influenced by their operational structure, and by their motivation to subsume ecological considerations into the overall enterprise. Dyllick has drawn attention to various reasons why green business measures can help producers to gain a competitive edge. Some of these are familiar, for example, the reduction of disposal problems, including those eventually faced by the consumer. Other considerations include the favouring of new product designs by eco-friendly segments of the market, the general corporate benefits of an environmentally conscientious image, the costs of reactive approaches to pollution legislation, a negative image associated with the manufacture and use of environmentally unfriendly products (e.g. CFCs, PET, non-returnable bottles), and the danger of building a business whose capital is the 'damaged' environment. A further, and perhaps more fundamental, factor is identified as the adoption of 'integrated' or 'holistic' concepts, which are claimed to reduce the risk of encountering liabilities. Hill (1992) has undertaken an illuminating review of the kinds of measures which have been incorporated by businesses in the pursuit of 'green' management, and these include a blend of audits, quality assurance systems, changes to management structures, in-house training, and targets.

Hopfenbeck (1993) considers a set of procedures, which he claims represent a 'holistic' basis for environmentally oriented business management. These can be summarised as:

- identification of the key ecological problems in the business;
- instilling ecological concepts in the values of management and staff, so that the overall culture is transformed;

Table 7.1   The Business Charter for Sustainable Development

Principles for environmental management

1. *Corporate priority*: To recognise environmental management as among the highest corporate priorities and as a key determinant to sustainable development; to establish policies, programmes and practices for conducting operations in an environmentally sound manner.

2. *Integrated management*: To integrate these policies, programmes and practices, fully into each business as an essential element of management in all its functions.

3. *Process of improvement*: To continue to improve corporate policies, programmes and environmental performance, taking into account technical developments, scientific understanding, consumer needs and community expectations, with legal regulations as a starting point; and to apply the same environmental criteria internationally.

4. *Employee education*: To educate, train and motivate employees to conduct their activities in an environmentally responsible manner.

5. *Prior assessment*: To assess environmental impacts before starting a new activity or project and before decommissioning a facility or leaving a site.

6. *Products and services*: To develop and provide products or services that have no undue environmental impact and are safe in their intended use, that are efficient in their consumption of energy and natural resources and that can be recycled, reused or disposed of safely.

7. *Customer advice*: To advise, and where relevant educate, customers, distributors and the public in the safe use, transportation, storage and disposal of products provided and to apply similar considerations to the provision of services.

8. *Facilities and operations*: To develop, design and operate facilities and conduct activities taking into consideration the efficient use of energy and materials, the sustainable use of renewable resources, the minimisation of adverse environmental impact and waste generation and the safe and responsible disposal of residual waste.

9. *Research*: To conduct or support research on the environmental impacts of raw materials, products, processes, emissions and wastes associated with the enterprise and on the means of minimising such adverse impacts.

10. *Precautionary approach*: To modify the manufacture, marketing or use of products or services or the conduct of activities, consistent with scientific and technical understanding, to prevent serious or irreversible environmental degradation.

11. *Contractors and suppliers*: To promote the adoption of these principles by contractors acting on behalf of the enterprise, encouraging and, where appropriate, requiring improvements in their practices to make them consistent with those of the enterprise; and to encourage the wider adoption of these principles by suppliers.

12. *Emergency preparedness*: To develop and maintain, where significant hazards exist, emergency preparedness plans in conjunction with the emergency services, relevant authorities and the local community, recognising potential transboundary impacts.

13. *Transfer of technology*: To contribute to the transfer of environmentally sound technology and management methods throughout the industrial and public sectors.

14. *Contributing to the common effort*: To contribute to the development of public policy and to business, governmental and intergovernmental programmes and educational initiatives that will enhance environmental awareness and protection.

15. *Openness to to concerns*: To foster openness and dialogue with employees and the public, anticipating and responding to their concerns about the potential hazards and impacts of operations, products, wastes or services, including those of transboundary or global significance.

16. *Compliance and reporting*: To measure environmental performance; to conduct regular environmental audits and assessments of compliance with company requirements, legal requirements and these principles; and periodically to provide appropriate information to the board of directors, shareholders, employees, the authorities and the public.

Source: ICC, 1991

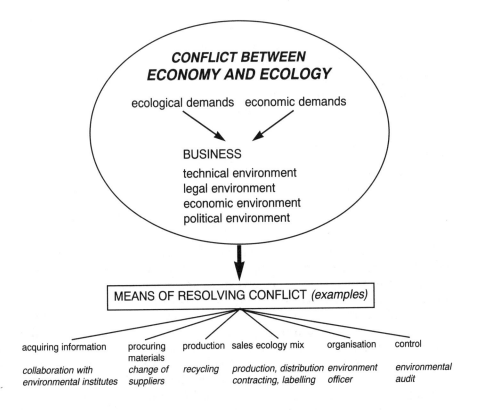

Figure 7.1   Elements of a comprehensive business environmental policy
Source: adapted from Meffert et al, 1987, cited in Hopfenbeck, 1993

- using results from current-status (i.e. baseline or state of environment) analysis to corroborate and, where necessary, amplify corporate environmental policy;
- drawing up an integrated system of objectives;
- determining adequate strategies;
- ensuring that 'green' considerations pervade every aspect of the business;
- ensuring that ecological thinking is embedded in institutional forms by nominating supervisors, environment committees, etc.; and
- installing a control system to ensure acceptable performance against environmental measures.

The elements of a comprehensive environmental policy for a business organisation are illustrated in Figure 7.1.

Clearly, organisational structure can strongly influence the scope for integrated management, and thus Roberts observes that changes to the organisation of a particular business function can provide both a stimulus for management innovation and an opportunity to tackle an old problem in a new way. He considers that traditional, hierarchical structures based on the rigid division of function into separate operating units has had an inhibiting effect on the introduction of integrated environmental management. These structures, it is argued, lead to poor communication and missed opportunities for materials recycling, energy management and purchasing policy. Further, employee isolation may exist and employees may have little influence on organisational management. This is contrary to:

> . . . the real message of sustainable business development, that is, that environmental problems can only successfully be resolved if all employees understand the causes and consequences of such problems, and if they are provided with the opportunity to take action in order to redress existing problems and to prevent the occurrence of environmental problems in the future.
>
> (Roberts, 1995, p.95)

In addition, complex industrial concerns are often pyramidal in structure, with individual cost centres being controlled from superior levels in the hierarchy. It has been argued that, in these circumstances, internal financial objectives can produce fragmented responses to a corporate environmental policy.

In general, it appears that the traditional hierarchical model of business organisation, as it is applied to large organisations, suffers from three major failings with regard to its ability to deal with sustainability. First, a rigid top-down system of policy, control and information can lead to inability to adapt to local environmental conditions. Second, the division into functionally specific and self-contained units of activity can hinder the adoption of holistic approaches. Third, the absence of flexibility to respond to changes in external conditions, such as new environmental regulations, can weaken the competitive position of divisional units and the company as a whole.

Some business theorists have therefore suggested that to be effective, the greening of a business must be coupled with more comprehensive organisational change, in which environmental issues are integrated with the periodic evolutionary changes that an organisation makes in order to respond to market conditions and remain competitive. To this effect, Gladwin (1993) offers various alternative models for organisational change, all of which provide opportunities for moving towards sustainable development. In the first option, greening is seen as a process of 'institutionalisation', in which the business organisation is reformed to allow it to conform to the environmental standards set for the organisation. Second, it may be associated with 'organisational learning', involving the acquisition of information and knowledge, so that the organisation 'learns' and this learning stimulates changes in behaviour. It may also reflect 'natural selection', in which organisations adapt in order to survive ('survival of the fittest' in a competitive situation) and where adaptation requires the transformation of organisational structures in response to external pressures. Alternatively, a 'strategic choice' model entails managers implementing new modes of organisation, products, processes and alliances. 'Transformational leadership' approaches require a proactive corporate attitude to achieve changes both in the organisational structure and in the values held by members of the organisation. Finally, change may be more 'evolutionary', in which organisations adapt through incremental change and innovation rather than waiting for frame-breaking change.

Corporate strategies for the environment

Businesses will be most effective in their environmental management if this is approached at corporate level, and the International Institute for Sustainable Development (1992) has proposed some key principles on which this may take place. The principle of *accountability* requires that an enterprise should account for the full scope of its resource use, not only within the conventional 'internal' business environment, but also on the external world. Further, the strategy must remain within the *financial capacity* of the organisation – it must be affordable and it should help identify any opportunities to obtain financial benefits from the adoption of enhanced environmental practices. Also, it must be capable of responding to *customer pressure*, recognising the requirements of customers and final consumers, and the demand for more environmentally acceptable goods. It is also important for the strategy to identify *competitive opportunities* which arise from adopting enhanced environmental attitudes and values. Strategic planning should also be alert to changing *public policy*, namely, probable and possible rises in the legislative standards affecting environmental performance. Some companies may be heavily reliant on scarce environmental goods, and changes in this type of dependency may be in their *strategic self-interest*. Finally, few managers are well equipped to understand and manage environmental issues, so that training programmes would be necessary to raise *management awareness*.

It has similarly been argued by Roberts (1995) that a green business strategy should embody a number of fundamental elements. First amongst these, he argued, was the practice of *discriminating development*, namely, that businesses should be discriminating in the use of resources in order to minimise waste and to prevent environmental and ecological damage. Second, business should give preference to the use of renewable resources and local resources should be used wherever possible (*'conserving resources'*). Third, the *'four Rs'* – repair, reconditioning, re-use and recycling – should be given priority in order to reduce the consumption of resources. Fourth, the principle of *'creative work'* requires work to be organised in such a way as to make the fullest possible use of human abilities and to involve people in ensuring that activities are conducted in a sustainable manner. Fifth, although growth which consumes resources is subject to eventual limits, this does not apply to activities such as the arts, education and leisure, which do not consume excessive amounts of resources; thus *'non-material growth'* should be maximised. Finally, *self-directed personal investment* requires the creation of opportunities to allow investment to take place in activities which will support sustainability and which will serve the needs of individuals and communities.

Criticisms have been made of many corporate 'green' strategies, in that they have tended to make misleading or overstated claims, or else have been too vague to monitor. Consequently various techniques have been developed which can give rigour and testability to claims of environmental performance. The starting point is the introduction of an 'environmental policy statement', which sets out fundamental standards to which a business commits itself. This contains an introductory statement asserting an overall philosophy; a typical example is:

> . . . our organisation is committed to promoting policies and actions which benefit the environment. We believe that encouraging everyone at our site to couple environmental concerns with everyday working practice will enable us to maintain our site as a clean, healthy and pleasant place for both employees and customers.
>
> (Newcastle Airport, quoted in O'Brien and Gibbins, 1994)

This initial statement would then be followed by a number of aims seeking, for example, to protect the environment in the course of the company's activities, to promote awareness among employees, and to ensure that current and planned activities incorporate environmental protection. Finally, in this idealised version, a set of specific objectives might include compliance with laws and regulations, consideration of environmental concerns in all company activities, co-operation with local interests, and endeavouring to use resources effectively and efficiently. An actual example of a policy statement for a local branch of a transnational corporation is given in Table 7.2.

At a more detailed level, an environmental management system should be introduced, similar to the approach described for green auditing within local authorities (the business approach was, indeed, cribbed for this purpose).

Table 7.2 Environmental policy and environmental objectives for a local business

Environmental policy

The following principles shall govern all business practices in the manufacture, procurement, maintenance, reuse/recycling and disposal of products and related services.

The protection of the environment and the health and safety of our employees, customers and neighbours from unacceptable risks takes priority over economic considerations and will not be compromised.

Operations at this branch must be conducted in a manner that safeguards health, protects the environment, conserves valuable material and resources and minimises the risk of asset losses.

This branch of the company is committed to manufacturing products and designing processes to optimise resources utilisation and minimise environmental impact.

All operations and products will be, at a minimum, in full compliance with applicable governmental requirements and company standards. Where conflicts arise between government regulations and company requirements, the more stringent requirements will apply.

This branch is dedicated to continuous improvement of its performance in Environment, Health and Safety. Environmental objectives will be published annually.

This plant's Environment, Health and Safety Policy, and Environmental Objectives will be made available to interested parties upon request. Copies of the Policy will be sent to local authorities and libraries together with copies of the company's Environmental Performance Review.

The Branch's Environmental Objectives for 1996 comprise a **vital few** environmental objectives and **community interface** objectives, namely:

vital few
- to maintain and develop our Energy Management System;
- as part of the ongoing energy management and conservation programme, institute a boiler-house optimisation review with a view to increasing efficiency and reducing utility usage;
- to achieve a 5% year-on-year improvement in the re-use and recycling of materials and parts as well as optimising carcass re-use opportunities; and
- to achieve a 5% year on year reduction in the weight of waste materials sent to landfill; and

community interface
- setting up an analytical programme to evaluate emissions from the site;
- continuing to monitor noise levels at the site perimeter;
- continuing to monitor purity of water in the local streams; and
- carrying out an investigation into the provision of on-site effluent treatment in order to reduce the number of waste tanker movements to and from the site.

Source: Rank Xerox, 1995; the policies relate to the Gloucestershire branch (Mitcheldean) of the parent company (Rank Xerox)

The most widely used method in Britain is that of BS7750, which influences the procedures operating at all levels within a business. It draws on the framework set by BS5750, on quality systems management, which, in essence, sought to ensure a 'deep and wide' permeation of quality consciousness throughout an organisation. The principal alternative method is that introduced by the European Union in 1993 – the Eco-Management and Audit Scheme – which contains some differences but, according to most observers, is in practice very similar (CEC, 1993). The European scheme involves the establishment and implementation of environmental policies, programmes and management systems by companies, in relation to their sites. It also seeks the systematic, objective and periodic evaluation of the performance of such elements, and the provision of information on environmental performance to the public.

Within the audit cycle, attention is to be given especially to:

- atmospheric emissions;
- discharges to water or sewers;
- solid or other wastes, particularly hazardous wastes;
- contamination of land;
- use of land, fuels and energy, and other natural resources;
- heat pollution, noise, odour, dust, vibration and visual impact; and
- ecosystem impacts.

The British Standard, introduced in 1992, sought to enable companies to establish procedures to set an environmental policy and objectives, achieve compliance with them, and demonstrate that compliance to others (BSI, 1992). The notion of 'auditing', in terms of setting standards and meeting procedural requirements, lies at the heart of the system.

A further technique which has found widespread application is that of 'life-cycle' (or cradle-to-grave) analysis. This is a method of identifying the environmental impacts that occur over the whole life-cycle of a product from the winning of raw materials to the product's eventual disposal as waste (Clark et al, 1993).Thus, for example, the European eco-labelling scheme uses an 'indicative assessment matrix' to identify the different types of environmental impact resulting from a product's manufacture, distribution, use and disposal (Figure 7.3). One of the main benefits of this approach has been to provide firm evidence of the causes of environmental damage, and thus the points at which policy might most effectively be targeted. Thus, for some products, issues of packaging, transportation and recyclability may be almost irrelevant compared to the product's performance in use. Life-cycle analysis appears to have great value both in making everyday products less environmentally damaging, and in revising preconceptions about what is 'green' and what is not. For example:

> . . . the difference between the best-performing and the worst washing machine on the market in Britain today is such that, if you bought the worst today, dumped it tomorrow and bought the best, there would be an environmental benefit within a year. Surprisingly, long product life is not necessarily 'green'.
>
> (Clark et al, 1993)

product life cycle	supply (materials, etc.)	production	distribution	use	disposal
impact					
air contamination					
water contamination					
soil contamination					
waste					
energy consumption					
local habitat					

Figure 7.3   The matrix used for the life-cycle assessment of washing machines according to the European eco-labelling scheme
Source: Blowers (ed.) 1993

It was previously observed that accountability and financial capacity are important ingredients in the assessment of a green corporate strategy. Gray (1994) has researched extensively the ways in which accountancy techniques can be extended, within a business context, to reflect the full environmental costs of a company's performance. Although a respected accountant himself, he has claimed that accountants, as a separate 'tribe' may be an entirely redundant breed in a 'deep green' world, and asserts that environmentalism can rise from its position as an 'also' issue in accounting to a matter of central importance. In essence, he repeats the claim that the environment can no longer simply be treated, by economists and accountants, as an externality, since it is integral (and thus internal) to any project which purports to enhance quality of life.

Gray proposes the collection of various sets of social and economic data for ecological accounting, namely:

- *input data*, e.g., on physical and human resources used, environmental impacts and disturbances caused, and ethical data on purchasing and investment activities;
- *processing data*, e.g., on efficiency of resource use, accidents and control of employees, and
- *output data*, e.g., on pollution emissions, wastes arising and other stresses.

These, together with other data on compliance with standards and perhaps also with ethical codes, would form the basis for determining 'accountability and transparency'. Taken together with the widespread evidence (from surveys and interviews) that the public is sceptical about companies' claims of environmental performance, which are often shrouded by opaque information, the scope for green accounting would appear to be considerable. Although many of the key issues could not be reported directly in financial terms, this is seen as a potential benefit rather than a problem. Gray sees these methods as aiding participatory democracy, and anticipates that they will fulfil two main functions. First, they would keep organisational decision-makers informed of the extent to which their particular organisation was depleting the planet's natural capital. Second, they would keep society informed about the way its capital was being used, and whether or not it was being maintained.

The local dimension

Confrontation between business and 'greens' has led to a good deal of antagonism and mutual distrust, and a large part of the sustainable development process must be directed to building bridges at the local level. Business involvement in environmental ventures has frequently been associated with giving money through sponsorship arrangements. While this method of assistance is often feasible, the view from business is that there are many other ways in which help and support could be donated to sustainability programmes. Businesses, especially the small and medium size enterprises (SME) which dominate the local scene, do not have a bottomless purse but may yet be able to bring other skills and knowledge into local partnerships.

One initiative of particular relevance is the Environment City movement, which has been pioneered in Britain in four urban areas (Leicester, Middlesbrough, Leeds and Peterborough). This project is substantially reliant on industrial sponsorship and amongst its various facets is the promotion of collaboration between the business community and other local stakeholders. Whilst this has obvious attractions to 'green' organisations because of the opportunity for sponsorship, it is also intended that SMEs in the locality may benefit from environmental partnership arrangements and from the resources made available by larger companies. In Environment Cities and elsewhere, business forums have been set up, in which sustainable development options have been explored, for example, through 'working breakfasts' and other entrées into the business world.

As an example, the business forum associated with Leeds Environment City (LEBF, 1993) has established five broad objectives:

1. to raise the awareness of the business community with regard to environmental issues;
2. to act as a focal point for the collection, dissemination and exchange of information and actions relating to best practice in the environment, and to act as an interface with organisations and agencies responsible for environmental regulation;
3. to share and distribute expertise and resources and to channel advice which will enable positive environmental action to occur;

4. to develop a full programme of environmental projects based on part-nerships between business, government and other appropriate organisations;
5. to publicise the achievements of the Leeds business community in the environment field, both within and outside the city.

Although this is a strong basis for business involvement in the environment, more general progress has been rather disappointing. The United Nations Association (1995) has suggested three reasons why this may be the case, which may be summarised as:

- economic recession, making it harder to get agreement for long-term investments;
- the way in which overtly green businesses have been exposed to public scrutiny, possibly making others nervous about following suit; and
- the lack of clear leadership from within the business community.

However, the UNA does draw attention to some more promising signs of a number of local authorities working beyond merely involving businesses in discussion. As one example, they cite the Local Economy Working Group in Luton – one of six groups set up there for the Local Agenda 21 process – which has started providing support for small businesses, including seminars and pilot schemes. They also note that a significant fillip for business involve-ment in LA21 occurred during 1995.

To complement the business dimension of Agenda 21, the World Wide Fund for Nature has launched a 'Business Agenda 21' project. In the mid-1990s, they commenced work in four areas with local authorities, vol-untary groups and businesses (with a focus on SMEs), the aim being to build up task forces of businesses in each area supported by local universities and voluntary organisations. It is proposed to develop 'tool kits', action plans, and documentation of best practice on issues such as energy and transport. These will stress the 'bottom line' advantages to SMEs, and will devise a methodology for action, trialling and change.

If the fashion for green marketing appears to favour affluent purchasers, it is worth noting that a further aspect of local economic sustainability is the emergence of local trading arrangements. These are essentially intended to by-pass the main retail and service framework, and thus represent local exchange trading schemes (LETS), independent of the main market economy. Inherent in the idea is a 'closed loop' local economy, which aims to increase the value of money spent by passing it through a network of local traders and suppliers, rather than allowing it to be 'leaked away' by local traders and suppliers whose revenues go to distant headquarters. Thus, local people trade goods with others and, at each transaction, a local currency (generally non-cash) is used and the value of the goods and labour adds to the sum total of the local currency in circulation. The accumulated value of transactions can then, potentially, be retained within the community. Although this approach is felt by some to reflect an unrealistic and excessively radical stance, it is worth noting that some 'indicators' being developed at local government level refer to 'percentage of local demand for building materials met locally' and the 'percentage of income spent locally' (UNA, 1995). Lang (1994) has also emphasised the role that LETS can play in local sustainability exercises, by helping re-institute a 'moral economy' of trust and interdependence between trading partners.

Workplace education

The ways in which we understand how attitudes change, and the timescales over which they may be influenced, has altered radically over the past generation. We are now well aware of the limitations of marketing propaganda, even though this may influence short-term commercial decisions; longer-term attitudinal changes occur over a much more extended time period, and are much more complex in the way they are formed. Promoting a change in attitudes is commonly achieved via education, whether this takes place inside schools or elsewhere. After leaving school or university, adults can often most effectively be reached through the workplace. Here, attitudes can be influenced in relation to environmental citizenship generally and, more specifically, individuals' contribution to green business performance.

Thus, Roberts (1995) refers to 'environmental awareness and training programmes' within businesses, and proposes that these should relate to the individual needs of the business organisation. He notes that there is little point in adopting a training scheme which is excessively ambitious, and which cannot be delivered. Within the workplace, a particular opportunity can be presented by the preparation of an environmental action plan. This may be produced with the help of employees to exploit opportunities for greening operational practices. One example of an EAP is that proposed by the *environmental citizenship* programme in Canada, which sprang from the Federal Green Plan. This involves:

- making wise and efficient use of resources – material, energy and water; and
- minimising contamination of the environment.

It thus entails knowing about the issues, and being willing to make responsible decisions about them. For business, the Green Plan suggests that there are a number of considerations in relation to environmental citizenship. These refer to demonstrating the organisation's commitment to the environment, to developing an action plan accommodating the business goals of a participating organisation, and fostering environmental citizenship in the broadest sense. Whilst the primary aim, therefore, may be to improve business performance, there are likely to be wider educational and citizenship benefits. Employees may become more conscious of the need for environmentally responsible behaviour and of the ways in which they can contribute to this as individuals.

Overall, it is intended that the EAP should signify the commitment to the responsible management of industrial and commercial products throughout their lifecycle, and should demonstrate the full commitment of management and employers throughout an organisation. Within the context of the Canadian Green Plan, organisations are encouraged to 'sign on' by adopting a set of environmental citizenship principles, and using these principles to develop a suitable in-house programme.

Conclusion

Local sustainability requires the integration of environmental, social and economic considerations. This broadening of the sustainability agenda away from traditional 'green' issues, and into the mainstream concerns of economy and society, has been one of the distinguishing features of the Agenda 21 process. The growing pressures on the industrial and commercial sectors generally to accommodate environmental and ethical principles appear to provide a positive context for local businesses. The pincer effect caused by legislative requirements (actual or potential), consumer demands and 'chaining' of responsibility, is leaving businesses little option but to take sustainability seriously.

Whilst environmental programmes have traditionally been associated with specific 'green up' measures and conservation projects, Agenda 21 is shifting the balance significantly towards economic development. Businesses, once at the margins of such programmes, or at least sought out only in the context of donating financial sponsorship, are becoming major participants. The workplace is also set to become a principal forum in which environmental citizenship principles can be communicated to adults. Whilst many businesses may be reluctant converts, there can be little doubt that sustainability issues are permeating throughout the private sector. Indeed, in some instances, indications are that the superficial 'green veneer' adopted for publicity purposes appears increasingly to be supplanted by the adoption of sustainability as a basic principle of business strategy and accounting. This may well emerge as the most signal hallmark of maturity in the environmental movement.

8

Citizens and local sustainability

Introduction

Some of the environmental problems we face are genuinely global as, for example, the depletion of stratospheric ozone. In many instances, however, it is misleading to talk of 'world' environmental or population problems: more helpfully, we can think in terms of a whole series of national or local problems. These can often most effectively be addressed, partly or wholly, by local community-based actions. Unfortunately, this normally presupposes that physically located communities can easily be identified, and that individuals will readily act in the common good.

One of the features of contemporary western society has been the disintegration of local communities, bound together by common interests and closely linked to a particular place. This is a process connected with de-traditionalisation, or 'individualisation' (reflected in the writings of Ulrich Beck, e.g. Beck, 1992), and it leads people to disaffiliate from their former norms of morals, kinship and ethnicity. (Conversely, it might be argued that people now, subconsciously, conform to new conventions and belong to new tribes which offer a basis for structured communication and participation.)

Insofar as communities exist, they tend to be diffuse networks of people with particular interests and activities in common rather than a single, all-purpose cluster of people in a locality. It could also be argued that people participate less in communal activities and give less readily of their time and energies to 'worthy' local organisations than was previously the case and that their activities are more home- and self-centred. Also, there appears to be a general drift away from commonly held moral values, to a more relativistic interpretation of what constitutes socially acceptable or unacceptable behaviour. In total, this does not augur well for the kinds of collective, public-spirited actions which may be required in the pursuit of local sustainability. Indeed, Moore

(1994) has argued cogently that the success of municipal programmes for change is strongly related to the levels of 'civicness' in the community.

A key ingredient in local action must be the individual citizen. Broadly speaking, there are two styles of citizenship which have been identified. One emphasises the individual's rights as a citizen, which are free benefits from residing in a liberal country. The citizen needs do nothing more than consume these, and so the style of citizenship may be entirely 'passive': this interpretation is sometimes referred to as 'liberal-individualist'. The alternative viewpoint is that a citizen possesses certain moral duties and obligations, and these should be exercised through 'active' engagement in the community. Thus, citizens accept a certain sense of civil obligation to pursue, in some degree, a collective way of life: this style is sometimes called 'civic-republicanism'. In view of the fragmentation of western society, and its arguably more selfish lifestyle, it is not surprising that we have increasingly displayed 'civic sclerosis' (Alinski, 1972), and only a relatively few people are willing to act as community champions. This condition is further reinforced where private land-ownership patterns are dominant, and government 'experts' have an overbearing role, so that effective citizen involvement is stifled.

This is clearly at variance with the need to engage citizens actively and collectively in the pursuit of local sustainability, and thus some countries have instituted 'environmental citizenship' programmes. Indeed, Principle 10 of the Earth Charter (included in Agenda 21) states that 'environmental issues are best handled with the participation of all concerned citizens, at the relevant level'. This seemingly incontrovertible statement actually contains (at least) two significant assumptions: that governments will be willing to permit power to diffuse away from the centre (suggesting, perhaps, the existence of a 'participatory democracy'); and that citizens are inclined to work collectively and participate actively on environmental issues.

Local discourses on citizenship are assumed to be effective in leading to changes in attitudes, behaviour and lifestyles. Some active citizenship campaigns have been criticised as means by which government may exploit 'platoons' of willing volunteers to undertake local tasks which, arguably, should be publicly-funded – an approach pejoratively termed 'platoonism'. However, much of the resurgence of interest in citizenship has been sincerely regarded as the most effective means of achieving long-term sustainability goals. In this regard, environmentalists are faced with somehow re-creating local communities as a focus for collective action.

The changing context of citizenship

Active environmental citizenship is encouraged within a context associated with an allegedly 'new' environmental agenda. It also coincides with a 'new' politics, where the traditional party political distinctions command diminishing public interest. Whereas politics has traditionally been associated with a battle between 'right' and 'left', this historical cleavage of interests is widely seen to be less relevant in a complex society where people identify less, if at all, with a particular 'class'. Equally, the extensive experimentation with collectivist approaches on the one hand and private-choice measures on the other has over the past century produced a wealth of evidence about the workability or otherwise of political parties' cherished policies. Thus, traditional political parties now often occupy barely distinguishable ground on certain major issues.

However, it is widely acknowledged that Western citizens are increasingly concerned about new 'political' issues (in the sense of debates surrounding issues where there is an imbalanced power relationship which needs to be addressed by policy mechanisms). These cover topics such such as race, gender and environment. Although the major political parties have developed policies on these issues, it has been suggested that they 'all march to the same drum'. Thus, whereas debates on education, health and so forth, tend to divide along predictable party lines, the party political dimension is something of an irrelevance in relation to environmental issues. It is likely, therefore, that new political alliances will be formed, outside the traditional party political arena, on topics relevant to sustainable development.

The 'new era' of environmentalism is associated with various features, which appear to be making it an extremely durable political issue. Past commentators have suggested that the environment was subject to an 'issue attention-cycle', in which interest waxed and waned according to media coverage of specific events. Whilst there continues to be clear evidence of changing popular attitudes in response to media reporting, it now seems that the environment is too deeply embedded as a mainstream issue to fall far off the agenda.

McGrew (1993) argues that this more substantive change is reflected in generally heightened public concern for the environment, the emergence of green consumerism, a broad diffusion of green values, and increased state regulation of topics such as pollution and conservation. He identifies four aspects of environmental politics which characterise its current status. First, there appear to be new social cleavages (i.e. new divisions in civil society), including the emergence of a stratum of highly educated professionals concerned about the delivery of high quality services. One interpretation of their evident concern for environmental amenities is that they are less closely enmeshed in industrial production than other social strata, and have thus embraced environmentalism as an expression of their own values and interests. Second, the kinds of strategies which 'green' non-governmental organisations use to achieve their ends are becoming well-established as a part of modern politics. These include agenda-setting, lobbying politicians and other decision-makers, and using the media in ways which ensure that incidents and issues become 'politicised'. Third, environmentalism is now starting to draw out the fundamental contradiction in state policy, namely, that increasing demands both for growth and for protection may prove to be irreconcilable, and to move it centre stage. Finally, the globalisation of environmental issues and the impact of international treaties and protocols make it very difficult for individual countries to stand aloof from making key decisions about sustainable development.

Against this background, our understanding of the meanings and importance of 'citizenship' have also experienced renewal and reinterpretation. The concept derives principally from ancient Greek civil society which held that free men should enjoy certain rights and responsibilities. In time, the debate was renewed, and eighteenth-century theorists developed a notion of 'civil citizenship' which established the rights necessary for individual freedom, such as rights to property, personal liberty and justice. This was followed in the nineteenth century by 'political citizenship', a concept which encompassed the right to participate in political power. It is within these frameworks that our previous references to passive and active citizenship can be placed. In the mid-twentieth century, an extremely influential series of lectures by T H Marshall (see Marshall and Bottomore, 1992) described the emergence of 'social citizenship', in which citizens' rights of economic and social security could be expressed through the modern welfare state. This enabled a much fuller participation of all individuals in the community and it was widely seen as the culminating stage of citizenship. To a large extent, citizenship then fell out of fashion as an area of academic enquiry, as it was felt that Marshall had written the definitive account. However, there has been a remarkable recent renewal of interest in the debate, not least attributable to the impact of environmentalists.

On the one hand, authors such as van Steenbergen (1994) have described the emergence of 'environmental citizenship' as a key phenomenon, and a principal instrument of renewal in the citizenship debate. On the other, post-Marshall theorists have noted that, even in sophisticated welfare states, there is a persistent and even increasing 'underclass' whose effective participation is hampered by serious social and economic disadvantage. Collectively, modern citizenship theory is strongly influenced by ideas of addressing poverty and extending civil rights to non-human components of the biophysical environment. Both of these are, in a sense, concerned with empowering the powerless and they are profoundly relevant to sustainable development.

Empowerment

If part of the environmental problem is that too much power has been centralised in government or corporations, then this imbalance must somehow be redressed so that local communities and individual citizens are enabled to make meaningful responses on their own behalf. Certainly, there is evidence to suggest that many people are apathetic to environmental issues, and thus to personal behavioural responses, because they feel it is the responsibility of those in authority. At the same time, opinion surveys indicate a low level of public trust in the competence, honesty and ability of governments and businesses to manage environmental resources to the benefit of all. Cram and Richardson (1992) note that, as a general principle, there is little point in allowing local citizens to participate in a highly consultative forum in close proximity to local issues, if the forum is effectively operating according to mandates issued by central government.

The effective empowerment of communities and citizens is a complex and risky business, although the term is often used glibly. When transferred to the local level, power can be dangerous: reporting of community views may be made by unelected, self-appointed activists, or by locally-based organisations with memberships that reflect a narrow interest. Whilst it is fashionable to talk of 'empowerment' it is important to bear in mind that normal democratic processes, such as the councillor system, should be included rather than overridden.

Two modes of transference of power to citizens are available. The first, more superficial, option is to view the citizen as consumer. Here, preferences are expressed through the market-place, and choices and values are reflected in the products and services purchased. This is the basis on which green consumerism is founded and, whilst flawed, it has some validity. The citizen's power as a consumer is potent, and one which must be incorporated into any theory of environmental citizenship. However, green consumerism cannot by itself offer a complete solution as there is clearly only a limited willingness to purchase products such as washing up liquids in an explicitly 'green' form. By contrast, there is a much more widespread expectation that environmental responsibility should be inherent across a much broader range of products and services. Consumerism, though, is the principle upon which the approach of the *citizen's charter* is founded and which essentially equates citizenship with the right to expect a certain standard or level of performance. The contents of the citizen's charter on the environment have previously been reviewed (Chapter 5), and this, though admittedly mainly 'passive' insofar as the consumer is concerned, represents a sort of empowerment of individuals and communities.

The second option, in which citizens are associated with grass-roots activism and direct action against perceived injustices is a far more radical one. Rather than advocating acquiescent, law-abiding conduct, this concept identifies the 'good citizen' as someone who is prepared, ultimately, to engage in civil disobedience and protest. There is a tension in this interpretation, in that citizenship broadly carries with it an obligation to respect a country's laws and legal apparatus. In environmental terms, this radical option has most commonly been associated with access rights in the countryside, pollution and animal welfare. Predictably, when politicians refer to citizenship and the transference of power, they do not have this model in mind, but expect their 'platoons' of volunteers to confine themselves to uncontroversial projects. However, a satisfactory model of local environmental citizenship will have to balance law-abiding and public-spirited activities with civil protest, and normal local government processes with genuinely devolved models of power.

Some observers believe that this balance will more effectively be achieved where the 'frontiers of the state' are progressively rolled back and governments take a more 'hands off' approach. However, there is no simple formula for success, as this latter attitude is often held by the 'new right' (as in relation to the dismantling of state regulations) which may be highly antipathetic to radical environmentalism.

The nature of environmental citizenship

One of the causes of poor and unsustainable management of the environment has been our excessive expectation of our 'rights' as consumers. Moreover, our ethical codes have related almost exclusively to the rights of humans and we have tended to regard ourselves as special animals which are superior to the rest of nature. One dimension of contemporary environmental citizenship is that we should cultivate an environmental ethic, which guides our actions as citizens of a wider community of nature. In this, we would have a sense of duty or obligation to the survival of nature more generally, and must be aware of the kinds of 'rights' which might attach to non-human species and even the physico-chemical environment.

It is generally held that citizens' rights and duties cannot be understood without an appreciation of the concerns and pressures that have led to them. This is profoundly true of environmental citizenship and it is therefore important briefly to consider the factors which have led to the emergence of modern environmentalism. A starting point for many analysts is to consider the impact of the Judaeo-Christian tradition, with its notion of humanity's 'dominion' over creation. It also asserts that humans are special, possessing souls, and that other species thus have less sanctity. Indeed, in this tradition, the first humans were given the right to 'name' species, which implies an imperialistic relationship – just as 'discoverers' of new lands first 'named' them, generally prior to subjecting them to colonisation. It has been argued by some that this world view has led Western society to take an imperialistic view of nature and, at the same time, of 'less developed' countries. It is very debatable whether this potential for greed is unique to Western society and, in any event, the argument is readily countered by the 'stewardship' doctrine, which springs from the same religious tradition. Here, it is held that, whilst humans may indeed enjoy higher status, they also have nobler and enduring responsibilities to care for 'creation'.

Both of these positions are simplistic and inaccurate interpretations of dominion and stewardship, but they have been highly influential in the development of ecophilosophy. It has, for example, been widely argued that some non-Western communities, or 'first nations' within economically advanced countries (e.g. Inuit, Maori), display more harmonious associations with nature. Whilst the disciplines of cultural and historical ecology do demonstrate that we have something to learn from traditional natural resource management methods, adherents of this view are prone towards misplaced faith in the myth of the 'noble savage' and the 'happy peasant'. More generally, there are now a number of radical 'green' analyses which emphasise the rights of nature (biorights), and these potentially transform our understanding of human rights and duties.

A classic analysis by O'Riordan (1981) has distinguished between the philosophical extremes of 'technocentrism' and 'ecocentrism', characterised by confidence in human ability to overcome environmental constraints in the former, and deference to the complexities of nature in the latter. Technocentrism is divisible into those views which affirm the Earth's ability constantly to supply human needs and adjust to human-induced pressures; and those which acknowledge nature's limits, yet consider that, with proper management of environmental impacts, these limits should not be transgressed. Ecocentrism is broadly characterised by its rejection of anthropocentric views of the world and of confidence in the power of 'advanced' technology. One position is to accept that human endeavours should be 'environment-led', and that environmental considerations should permeate all areas of human activity, leading to generally more conservative resource use. The most radical ecocentrist position is that which effectively denies the superior status of humans and which sees us as no more than one component in an integral web of nature. This 'deep green' viewpoint thus requires that our activity patterns should conform to the overarching dynamic of nature, and that there should be a broad equality of rights between humans, and animate and inanimate nature. The discourse of rights and duties associated with ecocentrism is often referred to as 'bioethics'. Poised between ecocentrism and technocentrism is the more recent notion of 'ecological modernisation' which, arguably, contrasts with 'progressive modernisation', thus replacing the modernist proposal of social reform and economic growth with a more environmentally sensitive agenda. Ecological modernisation allies humanistic and scientific aspects of the environmental debate and, whilst it goes beyond the tokenism of light-green philosophy, it is more institutionally bound than radical ecocentrism.

Clearly, no simple reconciliation of these different positions which would enable us to identify a unique bundle of responsibilities and rights which might be associated with environmental citizenship is possible. However, van Steenbergen (1994) has identified a number of distinctive characteristics which help to clarify our position. He notes that the history of citizenship has been one of increasing inclusion: in ancient Athens, for example, only free men could be citizens, with slaves, women and foreigners being excluded. Whilst political citizenship has obviously extended civil rights more generally, a consequence of radical environmentalism has been to extend the categories of inclusion further, even beyond those of present humans. Thus, rights can be extended to people yet unborn – which links to the previously expounded principle of intergenerational equity – and to nature, and in particular, to animals. There is a complex ethical debate, which can be followed elsewhere (e.g. Anderson, 1993), about the ways in which we can justify certain types of rights to other species. It may even be argued (Dobson, 1995) that animals have a right to be protected against genetic manipulation, which would have profound consequences for commercial agriculture and forestry.

Table 8.1 Extracts from the Rio Declaration on Environment and Development

Principle 1
Human beings are at the centre of concerns for sustainable development . . .

Principle 2
States have . . . the sovereign right to exploit their own resources . . . and the responsibility to ensure that activities within their jurisdiction . . . do not damage . . . areas beyond [their] jurisdiction.

Principle 3
The right to development must be fulfilled so as to . . . meet the . . . needs of present and future generations 'equitably'.

Principle 4
. . . environmental protection shall constitute an integral part of the development process . . . .

Principle 5
All States and people shall co-operate in the essential task of eradicating poverty . . . .

Principle 6
The special situation . . . of developing countries, particularly the least developed and those most environmentally vulnerable, shall be given special priority . . . .

Principle 7
States shall co-operate in a spirit of global partnership to conserve . . . the Earth's ecosystem . . . States have common but differential responsibilities . . . .

Principle 8
. . . States should reduce and eliminate unsustainable patterns of production and consumption and promote appropriate demographic policies.

Principle 9
States should co-operate to strengthen endogenous capacity-building for sustainable development . . . .

Principle 10
Environmental issues are best handled with the participation of all concerned citizens, at the relevant level . . . States shall facilitate and encourage public awareness and participation . . . .

Principle 11
States shall enact effective environmental legislation . . . .

Principle 12
States should co-operate to promote a supportive and open international economic system . . . .

Principle 13
States shall develop national law regarding liability and compensation for the victims of . . . environmental damage . . . .

Principle 14
States should . . . co-operate to prevent or discourage the . . . transfer to other States of any activities or substances that cause severe environmental degradation . . . .

Principle 15
. . . the precautionary approach shall be widely applied by States . . . .

Principle 16
National authorities should endeavour to promote the internalisation of environmental costs . . . .

Principle 17
Environmental assessment . . . shall be undertaken for proposed activities . . . likely to have a significant adverse impact . . . .

Principle 18
States shall immediately notify other States of any natural disasters or other emergencies . . . .

Principle 19
States shall provide prior and timely notification . . . on activities that may have a significant adverse transboundary environmental effect . . . .

Principle 20
Women have a vital role in environmental management and development . . . .

Principle 21
The creativity, ideals and courage of the youth of the world should be mobilised to forge a global partnership . . . .

Principle 22
Indigenous people and their communities, and other local communities, have a vital role to play in environmental management and development . . . .

Principle 23
The environment and natural resources of people under oppression . . . shall be protected.

Principle 24
Warfare is inherently destructive of sustainable development . . . .

Principle 25
Peace, development and environmental protection are interdependent and indivisible.

Principle 26
States shall resolve all their environmental disputes peacefully . . . .

Principle 27
States and people shall co-operate in good faith and in a spirit of partnership to fulfil the principles embodied in this Declaration . . . .

Source: based on the original Declaration contained in Agenda 21

A further hallmark of environmental citizenship is its emphasis on responsibility, which goes well beyond traditional notions of public-spiritedness towards the needs of community or nation. In the past, citizenship has included more and more categories of people as they have achieved emancipation. This has partly been achieved by their own direct action (struggle) in defending their rights, and partly by support from enlightened people who were already emancipated (or 'empowered'). Non-human species and the physical living and non-living environment, however, do not have a voice and cannot defend their interests; it is thus the responsibility of environmental citizens to 'speak up for nature' and to work out a set of bioethical principles. The defence of citizens' entitlements has, in many countries, been enforced by a 'Bill of Rights', which ensures impartial treatment of powerful and powerless alike. An attempt to enshrine a 'Charter' for the 'Earth' was made at the UNCED Conference; its 27 principles are summarised in Table 8.1.

Finally, the speed of communication between different parts of the world and the increasing inter-dependence of world trade patterns have led us to talk of the 'global village'. This is also important in the context of the pervasive impact of human activities on the global environment and of our awareness of the integrity of planetary air, water and biological systems. Thus, in a sense, we are beginning to recognise our duties as 'global citizens', with affinities and responsibilities to all humanity. Whilst we are clearly not citizens of a global community in the strict sense – we do not possess the same legal systems or governments – we do share many interests and concerns in common. We must therefore be very much aware of the interconnectedness of the world's economic and environmental systems, and of the complex consequences of our 'ecological footprints' beyond our own community or nation. In the context of the issues raised in the Brundt-land Report and Agenda 21, the notion of global citizenship can be seen to be of the greatest significance for sustainable development.

The good environmental citizen

Perhaps the major reason for strengthening the role of environmental citizen-ship is the recognition of the importance of local good practice, including the ways in which individuals may contribute to the resolution of global, interna-tional, national and local problems by 'acting locally'. The principle of good housekeeping – responsible action by individuals, households and commu-nities – and practical action is referred to as *primary environmental care* (PEC). This has been defined in *Caring for the Earth* as 'the process by which commu-nities organise themselves, strengthen their capabilities for environmental care, and apply them in ways that also satisfy their social and economic needs . . . .' (International Union for the Conservation of Nature and Natural Resources, 1991, p.57).

Clearly, the mechanisms of what is understood by PEC will vary substan-tially from place to place. Supported mainly by 'third world' examples, *Caring for the Earth* cites a number of actions to be accomplished in order to achieve PEC. These broadly entail:

- providing communities and individuals with secure access to resources and an equitable share in managing them;
- improving exchange of information, skills and technologies;
- enhancing participation in conservation and development;
- developing more effective local governments;
- caring for the local environment in every community; and
- providing financial and technological support to community environmental action.

It is assumed that these priority actions should together ensure that communities gain greater control over their own lives and can secure adequate resources, participate in decisions, acquire sufficient training and education, and identify means of meeting local needs in sustainable ways.

Table 8.2   Topics for citizen action

- protecting water resources;
- keeping toxic substances out of the environment;
- reducing smog;
- cutting waste;
- protecting natural areas;
- sustaining the diversity of wildlife;
- protecting historical heritage;
- stabilising greenhouse gas emissions;
- phasing out ozone-depleting substances;
- capping acid rain emissions; and
- sustaining renewable resources.

Source: Environment Canada, 1993

Some of these actions assume that effective local government, education and service delivery systems will not necessarily be in place and are thus more appropriate to circumstances found in less developed economies. Notwithstanding this caveat, the sentiments and some of the actions can be transferred to more affluent post-industrial states. Thus, the Countryside Commission (1993) has stated that:

> Individuals and communities at a local level are well placed to assess the needs of and to conserve their local countryside. Such 'primary environmental care' is fundamental to sustainability . . . . If people are enabled to respond to local environmental imperatives, they are more likely to grasp the issues of conservation and to move towards more sustainable lifestyles.

Once more, there is a supposed link between practising 'good housekeeping' as a process of self-education, and the progressive routinisation of responsible actions and expectations about environmental resource management more generally. As a broad generalisation, it could be argued that PEC is important as a device for overcoming civic sclerosis in developed countries, and for achieving indigenous low-to-intermediate technology solutions in less developed countries.

The practical interpretation of what PEC might entail has been attempted by some countries' environmental citizenship (EC) programmes. These programmes are examples of 'capacity building', since they entail processes which help people and organisations develop the skills necessary to manage their environment and develop it in a sustainable manner. The Canadian EC Program, for example, has involved the production of a number of 'primers' on key topics aimed at raising initial levels of awareness (Selman, 1994b). Emphasis is placed on the importance of civic responsibility towards the environment, supported by scientific reasoning, but also leading on to a whole range of practical actions that can be adopted in homes, schools and workplaces. It is argued that attainment of sustainable development is dependent on two key principles:

- *interdependence* between economic, ecological, social and cultural systems; and
- *participation* in decision-making, seeking a debate which is comprehensive in scope and in which all stakeholders are represented.

The primers identify over a hundred practical actions which individuals and organisations can undertake, grouped together under eleven broad headings (Table 8.2).

Against this background citizens are encouraged to become more actively involved. This will entail acquiring a better understanding of the environment, translating knowledge into responsible action, and using this knowledge and experience to enable and empower others. Environmental citizens (who, perhaps, equate with 'primary environmental carers') are thus expected to change their everyday habits, be responsible consumers, engage in public debate, keep elected officials accountable, and work with others.

One of the main components of environmental citizenship is that of education and, within it, the notion of environmental 'literacy'. Thus, a pre-requisite for effective participation and action is knowledge and comprehension of relevant issues and an aptitude to perceive ways in which the *status quo* might be changed for the better. It has been argued that the eco-literate citizen should:

> . . . have a blend of ecological sensitivity, moral maturity and informed awareness of natural processes at either individual or corporate levels
>
> (Brennan, 1994)

Although not all citizens could truly attain this level of competence, it does set a benchmark for the environmental sustainability elements of LA21s. It also contains two warning signals. First, whilst in a pragmatic fashion we may adopt an educative information-led agenda for cultivating local environmental responsibility, we may also need to embark on a more philosophical and fundamental reappraisal of our approach to the whole business of learning. Brennan goes as far as to claim that it is the western scientific tradition itself which lies at the heart of unsustainable development. Its excessively specialised, reductionist, disciplinary, framework-based style of enquiry is, he argues, at variance with a wider appreciation of the limitations of scientific knowledge and of the interconnectedness of natural resource systems. Second, we have previously noted that active citizenship may involve protest, dissent, and even illegal direct action. Policy-makers, in encouraging PEC, may find that they engage on a potentially risky and subversive business: but this is a necessary risk to take, and not one which should be evaded in favour of a hollow promise of 'empowerment'.

Activating citizens

Much of the rhetoric surrounding Local Agenda 21 has recognised that, if it is really to provide a framework for change, it must be propelled from the grass roots level. Thus, it has been argued (Golding, 1994) that the practice of subsidiarity needs to be embodied in a clearly linked partnership between institutions, national and local governments. However, few municipalities have countenanced the radical measures of community empowerment which may be necessary to give it full effect. The conventional (though, by previous standards, still quite radical) position is that local government should remain at arm's-length, without actually transferring political power. For example, LGMB (Stewart and Hams, 1991) has stated that '. . . devolution and de-centralisation does not mean autonomy. They will normaly take place within a framework set by the Authority'. This applies to the various extant experi-ments in decentralisation, associated with housing and health care.

Active environmental citizenship is a 'Pandora's box' so far as central and local governments are concerned and the extent to which it could either strengthen or undermine them cannot readily be predicted. One of the greatest tensions for LA21 will surely be that of balancing the need for political legit-imacy (afforded by local authorities) with the need to transfer real power to local communities and enable them to experience genuine and durable en-vironmental improvements. Golding (1994) argues that radical political action may need to be combined with executive support from local government of-ficers to break the mould of traditional thinking and establish new political and administrative models of government which truly empower the citizen.

One model which has been suggested for striking this delicate balance lies in the changing role of local authorities to facilitators rather than direct providers of most services. Some researchers have argued that the purchaser-contractor split, which has become the norm in local authority service provision, can be seen as an opportunity rather than a threat in this situation. It may be that this arrangement will enable local authorities to reassert their 'strategic leadership' role by concentrating on their overall policy direction, by virtue of their role as 'quality assessor' of contracted out services. Within this framework, it may become possible for many management and project functions to be devolved to communities. Local government could then concern itself with overall strategy, quality and performance, whilst neighbourhoods might achieve greater re-sponsibility for the implementation of particular programmes. This viewpoint would require mutual respect, and a partnership arrangement in which both sides accepted their respective responsibilities and the overall framework for making decisions. Equally, there are trends which are tending to remove power from, rather than cascade it towards, local communities.

Whilst it is fashionable to promote primary environmental care and neighbourhood-based action, we must be aware that this trend is accompanied by a great deal of rhetoric, both from reformers and radicals. The benefits of wider participation are not as self-evident as they at first appear and participa-tive strategies must be addressed realistically. In broad terms, the benefits arising from participation, partnership and citizen empowerment include the potentially deep and wide penetration of information and awareness about environmental change (Carew-Reid et al., 1994). This can be useful in inform-ing the analysis both of the (quantitative) magnitude and distribution of change, and the signficance (social and political impact) of the change. Policy formulation may be aided by widespread debate, and policies should become more 'transparent', that is, comprehensible and with no hidden agendas. As a consequence, they may become more credible, and there may be a wider sense of 'ownership' amongst the officials, volunteers and individual citizens who are meant to be adopting them. The existence of a broad constituency of interested parties may also assist with subsequent implementation and monitoring programmes. Set against these benefits, are the inevitable charges of 'paralysis by analysis'. Wider participation involves greater time, and may result in slow progress which can lead to frustration amongst the public. There is, in fact, much evidence from surveys that the lay public see the responsibility for the analysis and solution of environmental problems as lying firmly with government. Participatory processes also entail a lot of patience, effort and money in setting up meetings, managing the overall process and reporting on outcomes. Ultimately, the whole process may collapse, especially if the sponsors (upon whom most environmental programmes will be dependent) require a relatively quick and visible 'product'.

Presently, there is a general assumption that citizens should become more active, both as individuals and as members of wider processes and networks. Indeed, governments – both central and local – are increasingly inclined to take significant risks in involving the wider public. However, those responsible for organising environmental citizenship programmes should be aware of the pitfalls of loss of clarity where too many people are involved, loss of momentum and control, the escalating complexity of local situations, and the danger of raising unattainable expectations amongst participants. Environmental citizenship is an essential ingredient of local sustainabilty programmes, but it is also shrouded by a good deal of meaningless rhetoric which needs to be demystified.

Conclusion

Many LA21 programmes have emphasised the role of the citizen and have advocated a role of participatory democracy and environmental citizenship. The reasons for this are entirely comprehensible and generally laudable. However, the likelihood of a major resurgence of active citizenship associated with sustainability should be treated cautiously, and facilitators should be alert to the risks of responsibilities being devolved into the hands of perennial local activists. Neighbourhood-based approaches need to be linked to long-established channels of representative democracy and, in particular, to the legitimation which is bestowed by the councillor system. Equally, elected members should not feel threatened by the downward transference of power associated with local sustainability programmes, but should nurture ways of effective collaborative working.

It is equally clear that the approach taken in many countries during the early 1990s, which effectively treats the citizen as consumer, only addresses part of the issue. Sustainability programmes must find some way of overcoming our deeply ingrained civic sclerosis, and persuading people not only to modify their own actions, but also to participate in partnerships and networks of wider community change. In view of the previous comments about the need to balance grass-roots activism with the need for representativeness, it is clear that passive citizenship is not an alternative to active citizenship, nor *vice versa*. Rather, the challenge of environmental citizenship programmes will be to couple 'green charters' with agendas for more radical change, and generally to raise the level of 'civicness' amongst the community.

9

Conclusion

There is no doubt that the concept of sustainability has invaded the sphere of environmental policy and planning with spectacular effect. As yet, however, there is relatively little evidence of awareness of the concept amongst the lay public, and its utility as an effective basis for local action still requires further proof. There is, indeed, a risk of public reaction against the idea that the 'environment' and 'green issues' form an insufficient and inadequate basis of concern. Professionals and non-governmental organisations have, after all, spent three decades trying to imprint the need for environmental responsibility on the social conscience. Yet, whilst so much effort has been expended in communicating to people the need to go beyond traditional 'green' issues, professionals themselves are now starting to tire of the term 'sustainable development'. Even before it has caught on amongst the public, policy-makers and planners are, confusingly, seeking to coin further terms.

In the context of the present discussion, therefore, we need to pose two crucial questions about the notion of local sustainability, namely:

- whether it has local relevance, and
- whether it can remain durable until well into the twenty-first century.

If these can both be answered in the affirmative, then the message must be to 'keep the faith', and to pursue sustainable development regardless of our own over-familiarity with it as an expression. Below are a number of tests against which local sustainability can be evaluated. They are probably not exhaustive, and the brief comments on them which follow are little more than passing observations. However, hopefully, they form a basis for continuing debate and critique.

These tests are:

- Is sustainable development a durable concept?
- Does sustainable development address the kinds of 'quality of life' issues that are important to people?

- Does an awareness of sustainable development lead us to be better environmental citizens?
- Does sustainable development encourage the pursuit of primary environmental care?
- Does sustainable development help planners, policy-makers and managers to produce useful and effective local strategies?
- Does sustainable development aid frameworks for communicating options which are comprehensible to the non-specialists?
- Does sustainable development aid business to make better decisions?
- Does sustainable development give us a cogent and transparent vocabulary to express urgent issues?
- Is sustainable development an agent for change in the way we live?
- Does sustainable development matter to politicians and key decision makers?
- Can sustainable development help reach the disaffected?

Is sustainable development a durable concept?
Although many professionals now wince at references to 'sustainability' there is little doubt that it has focused creative energy on a wide range of practical projects and theoretical issues. In particular, it has led to the production of a suite of powerful economic principles which have given a welcome clarity to the ways in which we think about the continuing availability of environmental resources. It is unlikely that cogent concepts such as 'futurity' and 'critical natural capital' will quickly be jettisoned; rather, it is probable that they will become even more rigorous and applicable.

Does sustainable development address the kinds of 'quality of life' issues that are important to people?
One of the weaknesses of the 'green movement' has been its focus on traditional areas of environmental concern, such as pollution and species protection. Whilst these are obviously very important issues with wider social ramifications, it is apparent that public concern for them is episodic. Surveys reveal that the public have a much more continuing interest in the incidence of crime, employment levels, housing conditions, health and so forth. By focusing on the interface between society, economy and environment, it is clear that sustainable development articulates with a wider spectrum of 'quality of life' concerns.

Does an awareness of sustainable development lead us to be better environmental citizens?
In theory, greater awareness of an issue should lead us to behave more responsibly in relation to it. This is a fundamental principle underlying environmental education and interpretation. In practice, it is more difficult to demonstrate any measurable public response, perhaps because people have so many conflicting interests and loyalties and environmental concerns are a key priority for

only a minority of the population. Moreover, the notion of environmental citizenship cannot be disentangled from much broader questions of civicness and responsibility within an impersonal and morally relativistic society. Sustainability is inextricably linked to participatory approaches, yet these can be difficult to engender and sustain in contemporary communities. Nevertheless, it is clear that it may take a generation for attitudinal and behavioural change to occur: there are now many encouraging signs of environmental awareness and responsible attitudes amongst the younger members of society. Government, however, may have to create conditions in which environmental responsibilities and duties can more readily be exercised by, for example, introducing incentives to encourage recycling. More ambitiously, issues of sustainability can encourage people to see themselves as 'global citizens', and to become more appreciative of their role within a web of international responsibilities.

Does sustainable development encourage the pursuit of primary environmental care?
There is a real danger that sustainable development will remain merely a governmental platitude; yet there is scope for it to be a practical basis for local action by individuals and communities. There is still a good deal of public-spiritedness and concern for the locality at community level, and this can readily be harnessed to actions for sustainable development. Local communities, though, must feel that they are being genuinely supported (and even 'empowered') in their activity and that they are not merely being exploited as sources of volunteer labour to assist governments to achieve 'token' results.

Does sustainable development help planners, policy makers and managers to produce useful and effective local strategies?
Agenda 21 has been followed by a welter of Local Agenda 21s, some of which may to an extent merely amount to a re-badging of existing projects and programmes which may yield little more than novel rhetoric. However, to suggest that this applies to a majority of them would be a disservice to the enormous energy which many participants are injecting into the process at municipality level. Perhaps some of the initial progress on Local Agenda 21s has been disappointingly slow, but their organisers justify this in terms of the need to embed the necessary participatory and vision-building processes. Equally, many development plans in Britain are now being re-written with sustainability considerations to the fore, yet by the mid-1990s even these were only reaching the stage of formal 'adoption'. If sustainable development is to retain its credibility as a basis for effective local action, these plans and strategies must start delivering results which differ materially from traditional solutions.

Does sustainable development aid frameworks for communicating options which are comprehensible to the laity?
Traditional approaches to policy and planning have been 'top-down', and have tended to express choices in terms which are relatively abstract. They may be

easily assimilated by officers and politicians and even by well-informed voluntary organisations, yet they have rarely succeeded in engaging the wider public in a debate about future options. Sustainable development requires that challenging alternatives are posed to the public, who are then able to express clear preferences for particular courses of action. This is evidently a central concern to managers of Local Agenda 21 exercises, and there is some evidence of initial success.

Does sustainable development aid business to make better decisions?

There have been many assertions made by commentators on the business scene to the effect that 'green' business also makes commercial sense. This is a cosy assertion, but the reality is, of course, much more complex and the engagement of local businesses in environmental strategies and actions has, in quantitative terms, been rather disappointing. Nevertheless, the corporate policies and annual reports of some major companies and market leaders are very heartening: it is apparent that they have been deeply influenced by the introduction of 'sustainable development' as a touchstone of change.

Does sustainable development give us a cogent and transparent vocabulary to express urgent issues?

Sustainable development literature is peppered with terms such as 'empowerment', 'ownership' and 'vision'. These are concepts which have been used widely in the business world, where there is now often a cynicism about their practical meaning for the average employee. Also, they mean hugely different things to different communities, especially as between the developed and developing world. However, it is clear that, when allied to genuine organisational reform and adaptation, these practices have been spectacularly successful in facilitating effective change. The language of sustainable development places a strong emphasis on devolution of action and responsibility. Sincerely grounded, it will lead to powerful expectations about the ways in which change must be expressed and pursued.

Is sustainable development an agent for change in the way we live?

If society is to become more 'sustainable', we will need to make major alterations to the nature of our governance, personal behaviour, ethical bases for action and ability to cope with uncertainty. One practical way in which sustainable development is already helping us to do this is by introducing, or being inserted into, an ethos of continuous improvement in working practices. Properly instilled, the spirit of quality assurance can help us comprehend and cope with the pressures for rapid change.

Does sustainable development matter to politicians and key decision makers?

If sustainable development does not really matter to politicians and decision-makers, then it will fall off their (and perhaps the) agenda. Presently, there is doubtless a degree of tokenism and transience in government pronouncements on sustainability. However, sustainability is now so embedded in high level treaties and other legal and quasi-legal documents, that it will be impossible for governments rapidly or respectably to abandon the concept. Perhaps more probably, measures of sustainability and 'quality of life' will start to be used by politicians as an alternative to traditional indices of economic growth, as the latter are likely only to display sluggish and unimpressive results.

Can sustainable development help reach the disaffected?

Of all the challenges to the pursuit of local sustainability, this is probably the most difficult. Locally-based projects are, by themselves, most unlikely to overcome the alienation which exists within modern, complex societies. However, political rhetoric is now starting to embrace the need for renewal of active citizenship as a desirable aim, and to address the real problems associated with a growing fringe of disaffected people. It is highly probable that sustainable development will make a major contribution to this goal.

These reflections are only a few of the many which could have been made. They are enough, it is hoped, to affirm the need to adhere to sustainable development as a powerful and durable framework for social progress into the next century and to dissuade professionals from tiring of the term just as it is starting to reach the lay public. Sustainable development has captured the international imagination and unified collaborative endeavours to a rare degree: it is now well poised to start delivering real results which will be of benefit to people in their local communities.

Bibliography

Alinski, S. (1972) *Rules for Radicals*, Random House, New York.

Anderson, J.C. (1993) Species equality and the foundations of moral theory, *Environmental Values*, 2, 347–65.

Arnstein, S. R. (1969) A ladder of citizen participation, *Journal of the American Institute of Planners*, 35, 216–224.

Aygeman, J. and Evans, B. (eds.) (1994) *Local Environmental Policies and Strategies*. Longman, Harlow.

B&Q plc (1995) *How Green is My Front Door?* B&Q's Second Environmental Review. B&Q plc Quality and Environmental Department, Eastleigh.

Barton, H. and Bruder, N. (1995) *A Guide to Local Environmental Auditing*. Earthscan, London.

Bayfield, N. and McGowan, G. (1995) 'Monitoring and managing impacts on landscape: a case study of the Aonach Mor ski resort 1988–1995', in Griffiths, G.H. (ed.) *Landscape Ecology: theory and application*, pp. 93–101. IALE-UK, Reading.

Beck, U. (1992) *Risk Society: Towards a New Modernity*, Sage, London.

Beer, A. (1994) 'Developing tools to monitor the effectiveness of development plans', in van der Vegt, H. et al., (eds.), 69–88.

Blowers, A. (ed.) (1993a) *Planning for a Sustainable Environment*. A Report by the Town and Country Planning Association. Earthscan, London.

Blowers, A. (1993b) The time for change, 1–18, in Blowers (ed) (1993a).

Boothroyd, P., Green, L., Hertzman, C., Lynam, J., McIntosh, J., Rees, W., Manson-Singer, S., Wackernagel, M. and Woollard, R. (1994) Tools for Sustainability: Iteration and Implementation. In Chu, C.M. and Simpson, R. (Eds.) *Ecological Public Health – From Vision to Practice*, Centre for Health Promotion, University of Toronto, 111–121.

Borgstrom, G. (1967) *The Hungry Planet*. Collier, New York.

Bosworth, T. (1993) 'Local authorities and sustainable development', *European Environment*, 3(1), 13–17.

Boucher, S. and Whatmore, S. (1993) Green gains? Planning by agreement and nature conservation, *Journal of Environmental Planning and Management*, 36, 33–50.

Breheny, M. (ed.) (1992) *Sustainable Development and Urban Form*. European Research in Regional Science, no. 2. Pion, London.

Breheny, M. and Rookwood, R. (1993) Planning the sustainable city region, in Blowers, A. (ed.), 150–189.

Brennan, A. (1994) Environmental literacy and educational ideal, *Environmental Values*, 3, 3–16.

British Standards Institute (1987) BS5750: Quality Systems, Part 1: *Specification for design/development, production, installation and servicing*. BSI, Milton Keynes.

BSI (1991) BS5750: Part 8: *Guide to quality management and quality systems elements for services*. BSI, Milton Keynes.

British Standards Institute (1992) BS7750: *Specification for environmental management systems*. BSI, Milton Keynes.

Brugman, J. (1994) Sustainability Indicators: do we need them? International Council for Local Environmental Initiatives *Newsletter*. p.1 and 12.

Carew-Reid, J., Prescott-Allen, R., Bass, S. and Dalal-Clayton, B. (1994) *Strategies for National Sustainable Development: a Handbook for their Planning and Implementation*. IUCN/IIED, Earthscan, London.

Central and Local Government Environment Forum Working Group (1994) *Your Council and the Environment: The Model Local Environment Charter*. DoE, London.

Clark, M., Burall, P. and Roberts, P. (1993) A sustainable economy, in Blowers, A. (ed), 131–149.

Commission of the European Communities (1990a) *Europe 2000: Outlook for the Development of the Community's Territory*. COM (90) 544. CEC, Brussels.

Commission of the European Communities (1990b) *Green Paper on the Urban Environment*. CEC, Luxembourg.

Commission of the European Communities (1992) *Towards Sustainability – A European Community Programme of Policy and Action in Relation to the Environment and Sustainable Development*. COM 92 (23). CEC, Brussels.

Commission for the European Communities (1993) Council Regulation (EEC) No. 1836/93 of 29 June 1993 allowing voluntary participation by companies in the industrial sector in a Community eco-management and audit scheme. *Official Journal* I, 168, Vol 36.

Countryside Commission (1993) *Sustainability and the English Countryside: Position Statement*. The Countryside Commission, Cheltenham.

Cram, L. and Richardson, J. (1992) *Citizenship and Local Democracy: a European Perspective*. Report for the Local Government Management Board. LGMB, Luton.

Danish Energy Agency (1993) *Energy Efficiency in Denmark*. DEA, Copenhagen.

Davis, P. (1994) 'Eco-management and auditing in local government', *Environmental Policy and Practice*, 3(4), 287–292.

Department of Environment (1992a) *Development Plans and Regional Planning Guidance*. HMSO, London.

Department of Environment (1992b) *The Countryside and the Rural Economy*. Planning Policy Guidance Note No. 7, HMSO, London.

Department of Environment (1993a) *Reducing Transport Emissions Through Planning*. HMSO, London.

Department of Environment (1993b) *Environmental Assessment of Development Plans: A Good Practice Guide*. HMSO, London.

Department of Environment (1994a) *Transport* (Planning Policy Guidance Note 13). HMSO, London.

Department of Environment (1994b) *Partnerships in Practice*. DoE, London.

Department of Environment (1995) *Guide on Preparing Environmental Statements for Planning Projects.* HMSO, London.

Department of Environment (1996) *Mineral Planning Guidance Note 1: General Considerations and the Development Plan System (Revised).* HMSO, London.

Department of Environment/Ministry of Agriculture, Fisheries and Food (1995) *Rural England: A Nation Committed to a Living Countryside.* ('The Rural White Paper). HMSO, London.

Dobson, A. (1995) Biocentrism and genetic engineering, *Environmental Values*, 4, 227–240.

Douglas, T. (1992) Patterns of land, water and air pollution by waste, 150–171, in Newson, M. (ed), *Managing the Human Impact on the Natural Environment*, Belhaven, London.

Dyllick, T. (1989) Ökologisch bewusste Unternehmungsfühtung: Der Beitrag der Managementlehre. Swiss Association for Ecologically Conscious Management, St. Gallen.

Emmelin, L. (1996) Landscape impact assessment: a systematic approach to landscape impacts of policy, *Landscape Research*, 21(1), 13–25.

English Tourist Board (1991) *Tourism and the Environment: Maintaining the Balance.* ETB, London.

Environment Canada (1993) *A Primer on Environmental Citizenship.* EC, Ottawa.

Environment Canada (1995) Sustaining Canada's Forests: timber harvesting, *State of Environment Bulletin* No. 95–4.

Environment Council (1995) *Environmental Resolve: Beyond Compromise: building consensus in environmental planning and decision-making.* Environment Council, London.

Fairlamb, C. (1994) 'Sustainable development', *Environmental Policy and Practice*, 4(1), 43–49.

Fischer, F. and Forester, J. (eds.) (1993) *The Argumentative Turn in Policy Analysis and Planning.* UCL Press, London.

Fisher, F. and Ury, W. (1991) *Getting to Yes: Negotiating an Agreement without Giving in.* (Revised edition). Business Books, London.

Forrester, S. (1990) *Business and Environmental Groups: a Natural Partnership?* Directory of Social Change, London.

Freeman, C., Littlewood, S. and Whitney, D. (1996) Local government and emerging models of participation in the Local Agenda 21 process, *Journal of Environmental Planning and Management*, 39(1), 65–78.

Gibbs, D. (1993) European environmental policy and local economic development, *European Environment* 3(5), 18–22.

Gibbs, D. and Healey, M. (1995) Local government, environmental policy and economic development, in Taylor, M. (ed) *Environmental Change: Industry, Power and Policy*, Avebury Press, Aldershot, 151–167.

Gilbert, M. (1994) BS7750 and the eco-management and audit regulation, *Eco-Management and Auditing*, 1(2), 6–10.

Gladwin, T.N. (1993) The meaning of greening: a plea for organisational theory. In Fischer, K. and Schot, J. (eds) *Environmental Strategies for Industry.* Island Press, Washington, DC.

Glasson, J., Therivel, R. and Chadwick, A. (1994) *Introduction to Environmental Impact Assessment.* UCL Press, London.

Golding, A. (1994) Empowerment and decentralisation, 101–114, in Aygeman, J and Evans, B.

Gordon, J. (1994) *Canadian Roundtables and Other Mechanisms for Sustainable Development in Canada*. Report for Local Government Management Board. LGMB, London.

Government of Canada (1991) *Report on Progress Towards a Set of Environmental Indicators*. Ottawa.

Government of Canada (1996) *The State of Canada's Environment*. Ottawa.

Gray, R.H. (1994) Corporate reporting for sustainable development: accounting for sustainability in 2000 AD. *Environmental Values*, 4, 17–45.

Grubb, M., Koch, M., Munson, A., Sullivan, F. and Thomson, K. (1993) *The Earth Summit Agreements: a Guide and Assessment*. Earthscan, London.

Gustafsson, G. (1994) 'Örebro', in van der Vegt, H. et al. (eds.), 97–104.

Hams, T., Jacobs, M., Levett, R., Lusser, H., Morphet, J. and Taylor, D. (1994) *Greening Your Local Authority*. Longman, Harlow.

Hardin, G. (1968) Tragedy of the commons, *Science*, 162, 1243–1248.

Haughton, G. and Hunter, C. (1994) *Sustainable Cities*. Regional Policy and Development Series No. 7. RSA, London.

ter Heide, H. and Berends, J. (1994) Guideposts to the ecological city, in Vegt et al, 119–131.

Her Majesty's Government (1994a) *Sustainable Development: the UK Strategy*. HMSO, London.

Her Majesty's Government (1994b) *Biodiversity: the UK Action Plan*. HMSO, London.

Her Majesty's Government (1994c) *Sustainable Forestry: the UK Programme*. HMSO, London.

Her Majesty's Government (1994d) *Climate Change: the UK Programme*. HMSO, London.

Hertfordshire County Council (1994) *County Structure Plan Review: Future Directions*. HCC, Hertford.

Hill, J. (1992) *Towards Good Environmental Practice: a Book of Case Studies*, Institute of Business Ethics, London.

HMSO (1993) *A Guide to the Eco-Management and Audit Scheme for UK Local Government*. HMSO, London.

HMSO (1996) *Indicators of Sustainable Development for the United Kingdom*. HMSO, London.

Holliday, J. (1993) Ecosystems and natural resources, 36–51, in Blowers, A.

Hopfenbeck, W. (1993) *The Green Management Revolution: Lessons in Environmental Excellence*. Prentice-Hall, London.

Hughes, R.V. (1994) Corporate funding for environmental projects, *Environmental Policy and Practice*, 4(1), 33–37.

Independent Commission on International Development Issues (Chair: W. Brandt) (1980) *North-South: a Programme for Survival*. Pan Books, London.

International Chamber of Commerce (1991) *Business Charter for Sustainable Development*. ICC, Paris.

International Institute for Environment and Development (1995) *Citizen Action to Lighten Britain's Ecological Footprints*. A report by IIED for the DoE. IIED, London.

International Institute for Sustainable Development (1992) *Business Strategy for Sustainable Development*. IISD, Winnipeg.

International Union for the Conservation of Nature and Natural Resources (1980) *World Conservation Strategy*, IUCN, Gland.

International Union for the Conservation of Nature and Natural Resources (World Conservation Union) (1991) *Caring for the Earth.* IUCN, Gland.

Jansson, A-M, Hammer, M., Folke, C. and Constanza, R. (eds.) (1994) *Investing in Natural Capital: the Ecological Economics Approach to Sustainability,* Island Press, Washington D.C.

Kirklees Metropolitan Borough Council (1989) *Kirklees: State of Environment Report* KMDC, Huddersfield.

Koslowski, J. and Hill, G. (eds.) (1993) *Towards Planning for Sustainable Development: a Guide for the Ultimate Environmental Threshold Method.* Avebury, Aldershot.

Krueger, R. (1994) *Focus Groups: a Practical Guide for Applied Research.* Sage, London.

Lancashire County Council (1991) *Lancashire: a Green Audit.* LCC, Preston.

Lang, P. (1994) *Rebuilding the Local Economy.* Grover Books, Bristol.

Lee, K. (1993) *Compass and Gyroscope: Integrating Science and Politics for the Environment.* Island Press, New York.

Leeds Environmental Business Forum (1993) *Good Environmental Business Practice Handbook.* LEBF, Leeds.

Leicester City Council (1995) *Blueprint for Leicester. Findings Report.* LCC, Leicester.

LGMB (1992a) *UK Local Government Declaration on Sustainable Development.* Produced by LGMB for the Local Authority Associations. LGMB, Luton.

LGMB (1992b) Earth Summit: Rio '92, Supplement No. 2, *Agenda 21 – a Guide for Local authorities in the UK.* Written for LGMB by CAG Consultants. LGMB, Luton.

Local Government Management Board (1994) *Greening Economic Development.* LGMB, Luton.

Lowe, P. and Murdoch, J. (1993) *Rural Sustainable Development.* Rural Development Commission, Salisbury.

Marshall, T.H. and Bottomore, T. (1992) *Citizenship and Social Class* (The Marshall Lectures, Cambridge 1949, reprinted with commentary). Pluto Press, London.

Martin, S. (1995) Partnerships for local environmental action: observations on the first two years of Rural Action for the Environment, *Journal of Environmental Planning and Management,* 38, 149–165.

Massam, B.H. (1988) *Environmental assessment in Canada: Theory and Practice.* Canada House Lecture Series No. 39, 32pp. Cited in Douglas (1992).

Masser, I. and Pritchard, A. (1994) Geographic information systems in state of the environment auditing, *Town Planning Review,* 65, 205–213.

McGrew, A. (1993) The political dynamics of the 'new environmentalism', in Smith, D. (ed.) (1993) *Business and the Environment: Implications of the New Environmentalism,* pp.12–26. Paul Chapman Publishing, London.

Macnaghten, P., Grove-White, R., Jacobs, M. and Wynne, B. (1995) *Public Perceptions and Sustainability in Lancashire – Indicators, Institutions, Participation.* A Report to Lancashire County Council by the Centre for the Study of Environmental Change, University of Lancaster.

Meffert, H., Benkenstein, M. and Schubert, F. (1987) Unweltschutz und Unternehmensverhalten, *Harvard Manager,* 2. Cited in Hopfenbeck (1993), p. 60.

Moore, J.L. (1994) *What's Stopping Sustainability? Examining the Barriers to Implementation of 'Clouds of Change'.* Summary of Master's Thesis, University of British Columbia, Vancouver.

Mormont, M (1996) Towards concerted river management in Belgium, *Journal of Environmental Planning and Management,* 39(1), 131–142.

Nijkamp, P., Lasschuit, P. and Soetman, F. (1992) Sustainable development in a regional system, in Breheny, M. (ed.) (1992), 39–66.

O'Brien, G. and Gibbins, C.N. (1994) Conducting a preliminary environmental review: a case study of Newcastle International Airport, *Environmental Policy and Practice*, 3(4), 273–286.

O'Callaghan, J.R. (1995) NELUP: an introduction, *Journal of Environmental Planning and Management*, 38, 5–19.

OECD (1990) *Environmental Policies for Cities in the 1990s*. OECD, Paris.

OECD (1991) *The State of the Environment*. OECD, Paris.

O'Riordan, T. (1981) *Environmentalism*. Pion, London.

O'Riordan, T. (1994) Civic science and global environmental change, *Scottish Geographical Magazine*, 110, 4–12.

O'Riordan, T. (1995) Managing the global commons, 347–360, in O'Riordan, T. (ed) *Environmental Science for Environmental Management*.

O'Riordan, T., Wood, C. and Shadrake, A. (1993) Landscapes for Tomorrow, *Journal of Environmental Planning and Management*, 36, 123–147.

Owens, S. (1992) 'Energy, environmental sustainability and land use planning', in Breheny, M. (ed.) (1992), 79–105.

Owen, S. (1996) Sustainability and rural settlement planning, *Planning Practice and Research*, 11(1), 37–47.

Pearce, D. (1993) *Blueprint 3: Measuring Sustainable Development*. Earthscan, London.

Pezzey, J. (1989) *Definitions of Sustainability*, UK Centre for Economic and Environmental Development, Cambridge.

Pickering, K. and Owen, L. (1994) *An Introduction to Global Environmental Issues*, Routledge, London.

Potter, C., Anderson, M. and Meaton, J. (1994) Playing the planner: public participation in the urban fringe, *Town and Country Planning*, 48–49.

Pritchard, D. (1994) 'Is carrying capacity real, and can we use it in planning? – an RSPB view', in Royal Society for the Protection of Birds (1994), *Proceedings of Annual Conference, Capacity Planning: a Practical Application of Sustainable Development in the Land-use Field*. 7–17. RSPB, Sandy.

Pritchard, S. (1995) UK Leads from Rio to Istanbul, *Planning Week*, 3(24), 13.

Raemaekers, J. (1993) Corporate environmental management in local government, *Planning Practice and Research*, 8, 5–13.

Rank Xerox Ltd. (1995) *Rank Xerox Environmental Performance Report, November 1995*. Rank Xerox, Marlow, Bucks.

Rees, W.E. (1992) Ecological footprints and appropriated carrying capacity: what urban economics leaves out, *Environment and Urbanisation*, 4, 121–130.

Richards, L. and Biddick, I. (1994) Sustainable economic development and environmental auditing: a local authority perspective, *Journal of Environmental Planning and Management*, 4, 487–494.

Roberts, P. (1995) *Environmentally Sustainable Business: a Local and Regional Perspective*. Paul Chapman Publishing, London.

Roebuck, S. and Gurney, A. (1995) 'Scope for growth', *Planning Week*, 3(22), 16–17.

Rural Development Commission (1993) Rural Sustainable Development. Written for the RDC by Lowe, P., and Murdoch, J. RDC, Salisbury.

Rural Development Commission (1995) *Planning for People and Prosperity. The RDC's Policy Statement on Planning in Rural England*. RDC, Salisbury.

SAFE (1994) *Food Miles Report*, SAFE Alliance, London.

Seddon, M.H. (1994) A green plan for Wakefield, *Environmental Policy and Practice*, 4(1), 39–42.

Selman, P. (1993) Landscape ecology and countryside planning: vision, theory and practice, *Journal of Rural Studies*, 9, 1–21.

Selman, P. (1994a) Canada's environmental citizens: innovation and partnership for sustainable development, *British Journal of Canadian Studies*, 9, 44–52.

Selman, P. (1994b) Systematic environmental reporting and planning: some lessons from Canada, *Journal of Environmental Planning and Management*, 37(4), 461–476.

Selman, P. (1995) Local sustainability: can the planning system help get us from here to there?, *Town Planning Review*, 66, 287–302.

Selman, P. (1996) The potential for landscape ecological planning in Britain. In: Curry, N. and Owen, S. (Eds.) *Changing Rural Policy in Britain: Planning, Administration, Agriculture and the Environment*, Countryside and Community Press, Cheltenham, 28–43.

Staffordshire County Council (1992) *Staffordshire County Council's Environmental Audit*, Parts 1 and 2. SCC, Stafford.

Stanners, D. and Bourdeau, P. (eds.) (1995) *Europe's Environment: the Dobříš Assessment*. European Environment Agency. Copenhagen, Earthscan.

van Steenbergen, B. (1994) Towards a global environmental citizen, 141–152, in van Steenbergen, B., *The Condition of Citizenship*, Sage, London.

Stewart, J. and Hams, T. (1991) *Local Government for Sustainable Development: the UK Local Government Agenda for the Earth Summit*. LGMB, Luton.

Sturt, A. (1992) Going Dutch, *Town and Country Planning*, 61(2), 48–51.

Thomas, H. and Tewdwr-Jones, M. (1995) Beacons Planners Get Real, *Planning*, 1104, 20–21.

Tjallingii, S.P. (1995) *Ecopolis – Strategies for Ecologically Sound Urban Development*. Backhuys, Leiden.

Trudgill, S. (1990) *Barriers to a Better Environment: What Stops Us Solving Environmental Problems?* Belhaven, London.

Tuininga, E-J. (1994) Going Dutch in environmental policies: a case of shared responsibility, *European Environment*, 4(4), 8–13.

Twine, F. (1994) *Citizenship and Social Rights – the Interdependence of Self and Society*, Sage, London.

United Nations Association UK (1995) *Towards Local Sustainability: a Review of Current Activity on Local Agenda 21 in the UK*. UNA, London.

United Nations Conference on Environment and Development (1992) *Agenda 21 – Action Plan for the Next Century*, UNCED, Rio de Janeiro.

van der Vegt, H., ter Heide, H., Tjallingii, S. and van Alphen, D. (eds.) (1994) *Sustainable Urban Development: Research and Experiments*. Proceedings of a PRO/ECE – workshop held in Dordrecht, November 1993. Delft University Press, Delft.

VROM (1993) *National Environmental Policy Plan 2*. Ministry of Housing, Spatial Planning and the Environment, the Hague.

Wackernagel and Rees, W. (1996) *Our Ecological Footprint: Reducing Human Impact on the Earth*. New Society Publishers, Gabriola Island, British Columbia.

Waddock, S.A. and Post, J.E. (1990) Catalytic alliances for social problem solving, 342–346, in Jauch, L.R. and Wall, J.L. (eds), *Academy of Management Best Papers Proceedings*, Vol. 190, San Francisco, California.

Ward, S. (1993) Thinking global, acting local? British local authorities and their environmental plans, *Environmental Politics*, 2(3), 453–478.

Whittaker, S. (Ed.) (1995) *An International Guide to Local Agenda 21. First Steps: Local Agenda 21 in Practice*. HMSO, London.

Winter, M (1996) *Rural Politics: Policies for Agriculture and the Environment*. Routledge, London.

Wood, D.J. and Gray, B. (1991) Collaborative alliances: moving from practice to theory, Part II, *Journal of Applied Behavioural Science*, 27, 139–162.

World Commission on Environment and Development (the Brundtland Report) (1987) *Our Common Future*, Oxford University Press, Oxford.

Index

UNIVERSITY OF WOLVERHAMPTON
LEARNING RESOURCES